Signals and Communication Technology

More information about this series at http://www.springer.com/series/4748

Pramode K. Verma · Mayssaa El Rifai
Kam Wai Clifford Chan

Multi-photon Quantum Secure Communication

Springer

Pramode K. Verma
School of Electrical
and Computer Engineering
University of Oklahoma
Norman, OK, USA

Kam Wai Clifford Chan
School of Electrical
and Computer Engineering
University of Oklahoma
Norman, OK, USA

Mayssaa El Rifai
School of Electrical
and Computer Engineering
University of Oklahoma
Norman, OK, USA

ISSN 1860-4862 ISSN 1860-4870 (electronic)
Signals and Communication Technology
ISBN 978-981-10-8617-5 ISBN 978-981-10-8618-2 (eBook)
https://doi.org/10.1007/978-981-10-8618-2

Library of Congress Control Number: 2018949888

This Springer imprint is published by the registered company Springer Nature Singapore Pte Ltd.
The registered company address is: 152 Beach Road, #21-01/04 Gateway East, Singapore 189721, Singapore

Preface

Information is the currency of the modern age. Security of information will continue to be of paramount importance in the foreseeable future. A practical way of transferring unconditionally secure information does not exist today. Quantum key distribution (QKD) technologies come close, but they too are unconditionally secure only to the extent key (in other words, random information) transfers are involved. In order, then, for unconditionally secure information to be transferred, one must resort to using the securely transferred keys as a one-time pad, and X-or them with the payload information.

This book explores alternative ways that can accomplish secure information transfer without the need for a quantum channel as in the case of QKD-based techniques. We do not claim that the techniques presented here lead to theoretical or unconditional security, although we believe it can come close to those based on QKD techniques. Except for an interesting technology presented in Chaps. 11 and 12, the techniques presented in this book do not need conventional encryption. Most of the work presented in this book has been practically realized, albeit in a laboratory environment. Our objective has been to offer a proof of concept rather than build a rugged instrument that can withstand the rigors of a commercial environment.

A word about contemporary encryption techniques: First, no encryption technology other than those based on a one-time pad has been shown to be provably secure. From a practical standpoint, however, techniques based on a one-way mathematical function do meet the security requirements of most applications if the computing power available to the intruder is within the currently anticipated limits of computing power. The mathematical function itself behind an encryption algorithm is considered acceptable if the computing effort associated with a proposed cryptanalytic attack is not less than the computing effort necessary for a brute force attack.

The encryption techniques presented in this book (except in Chap. 4 and the last two chapters) have the following things in common: Encryption is carried out in a streaming manner as data is generated. No prior exchange of keys is involved. To avoid man-in-the-middle attack, the communicating parties are, however, expected to have a common initialization vector which can be updated as frequently as desired. In the multistage protocol, Alice and Bob choose their respective keys themselves, separately and independently of each other, with no need to intercommunicate their keys. We can reduce the transmission penalty by reducing the multistage transmission to single-stage transmission. In this case, however, keys must be exchanged, but they can be updated frequently as a nonlinear function of the actual data exchanged and the initialization vector. Of course, the single-stage mechanics can revert to the multistage configuration obviating the need for key exchange, or for generating a fresh seed key, as often as desired.

It is the authors' hope that the work presented here will lead to the exploration of additional techniques that can deepen our understanding and help develop a wider arsenal of secure information transfer instruments that can be applied to a variety of emerging scenarios in a practically realizable manner. A brief synopsis of the chapters in this book is as follows.

Chapter 1 of the book presents a general introduction to cryptography including its historical evolution over the past couple of thousand years. The chapter concludes by addressing the shortcomings of cryptography as practiced today and points to the need for introducing additional techniques that can withstand the conflicting demands of simplicity of realization and increasing cryptographic strength. In particular, it points to the need for the use of quantum mechanics based techniques in cryptography.

Chapter 2 gives the mathematical background of quantum mechanics used in the rest of the book. The abstract concept of a qubit as the quantum extension of a classical bit is first introduced. Characteristics of photons are then covered to lay the foundation for multi-photon communication. An exposition of the polarization degree of freedom of photons in the multi-photon regime is made.

Chapter 3 of the book offers a discussion of quantum key distribution techniques as practiced today along with their strengths and limitations. Protocols like BB84 and the related techniques, such as E91, B92, SARG04, and decoy states, are covered in this chapter.

Chapter 4 discusses a class of quantum communication protocols called KCQ that exploits the inherent quantum noise in measurement to protect information in transit. The KCQ protocol generally permits multiple photons in a signal pulse. A particular realization of KCQ, the widely reported Y-00 protocol, is discussed. It offers a convenient introduction to the rest of the book because the additional techniques presented in the book are also based on multi-photon technology.

Chapter 5 introduces the multi-photon three-stage protocol for realizing security without the need for conventional cryptography as necessary accompaniment for implementing QKD-based encryption techniques. The chapter describes the realization of the three-stage multi-photon protocol in free-space optics.

Chapter 6 generalizes the three-stage protocol into a family of multistage protocols. It compares the multistage protocol with single-photon protocols and illustrates how a multi-photon protocol can be made secure against man-in-the-middle attack. Since a multi-photon protocol is, in general, subject to photon-siphoning attacks, the protocol introduces another variable to thwart such attacks.

Chapter 7 presents a security analysis of the multistage protocol assessing its vulnerability to known security attacks. It shows that the multistage protocol can offer quantum level security under certain conditions.

Chapter 8 analyzes intercept-and-resend and photon number splitting attacks in the multistage multi-photon protocol. It lays down the conditions under which the multistage multi-photon protocol can approach the strength of a quantum-secure protocol.

Chapter 9 extends the application space of the multistage multi-photon protocol to wireless communication. It examines the viability of using the multistage multi-photon protocol for secure key distribution in the IEEE 802.11i protocol.

Chapter 10 presents a unique way of using the polarization channel of a fiber optic cable to detect the presence of an intruder. This layer-1 based intrusion detection system prohibits an adversary from capturing any information flowing on the cable.

In Chap. 11, we use the polarization channel to transfer keys to encrypt any channel on the fiber optic cable using conventional symmetric cryptography. The novelty lies in using the polarization channel as a convenient way to securely transfer symmetric encryption keys among the communicating parties.

Chapter 12 extends conventional cryptographic techniques to offer an ultra-secure router-to-router key exchange system based on the multistage protocol. The routers can be connected through a range of diverse transmission media.

Norman, USA
May 2018

Pramode K. Verma
Mayssaa El Rifai
Kam Wai Clifford Chan

Acknowledgements

This book is the outcome of collaborative effort among many individuals associated with the Quantum Optics Laboratory of the University of Oklahoma—Tulsa, and from those associated with other universities and institutions.

The authors would like to thank Dr. Subhash Kak from Oklahoma State University for his seminal work on the three-stage protocol that inspired them to explore this territory. Dr. Kak and Dr. Yuhua Chen from the University of Houston have participated in several discussions over the past 10 years during our investigation. Dr. Gregory MacDonald's doctoral work and his continuing collaboration on the use of the polarization channel as a communication medium has helped us refine our approach to make its best use for cryptography. Dr. Robert Huck has offered deep insight into all experimental work carried out in the laboratory. Without Dr. Huck's guidance and support, much of our work would have remained unexplored. The support of Dr. James J. Sluss, Jr., throughout these investigations and especially in equipping the Quantum Optics Lab is gratefully acknowledged.

Several students received their Master's and doctoral degrees based on their research in the Quantum Optics Laboratory. Much of this book is based on their published works—they form the backbone of this book. The authors are grateful to Shweta Bhosale, Bhagyashri Darunkar, Nilambari Gawand, Rasha El Hajj, Sayonnha Mandal, Rupesh Nomula, Nishaal Parmar, Nikhil Punekar, Mitun Talukder, Farnaz Zamani, and Lu Zhang, who led many investigations related to their research. The outcome of their research reflects throughout this book.

Pramode Verma would like to thank his wife Gita for her support during the preparation of the book, and especially for singlehandedly assuming the burden of our physical relocation while this book was work-in-progress. Mayssaa El Rifai would like to thank her beloved family: her dad Jihad, mom Maha, sisters Rihab and Riham, husband Samer, and daughter Rita for their encouragement and support during the writing phase of this book. Kam Wai Chan would like to thank his wife Chung Ki for her support during the preparation of this book as well as throughout the years.

Contents

List of Figures

Chapter 1
Introduction

This chapter offers a brief history of cryptography and reviews the classical and contemporary methods of securing information.

1.1 Cryptography

> *The multiple human needs and desires that demand privacy among two or more people in the midst of social life must inevitably lead to cryptography wherever men thrive and wherever they write.*
>
> —David Kahn

Cryptography is the art of secret writing. It encompasses the field of applications that provide authentication, privacy, integrity and confidentiality to users. Cryptography has performed an important role in the history of any society that depends on information [1]. An important subfield of cryptography is that of secure communication. This field aims at protecting any message during the process of its transfer between communicating parties such that no unauthorized party can meaningfully access the content of a message in transit. This book is about protecting the confidentiality and integrity of information in transit.

Contemporary cryptography is the process of transforming digital information into a sequence of bits which is incomprehensible to anyone other than the intended user. The process of transformation is referred to as encryption; the reverse process is referred to as decryption. The information or message to be encrypted is referred to as plaintext; after encryption, this information is called a ciphertext. Over an extended period of time covering several centuries, many methods to encode (or encrypt) messages have emerged, always to be broken at a later point in time.

Cryptology, which covers both encryption and decryption, can still be considered a young science. Even though cryptography has been used for about two thousand years as a way to protect messages, its systematic study as a science did

P. K. Verma et al., *Multi-photon Quantum Secure Communication*,
Signals and Communication Technology,
https://doi.org/10.1007/978-981-10-8618-2_1

not begin until around a hundred years ago. In the next section, a brief history of cryptography is offered starting from the first known evidence of its usage in Egypt up until now.

The concept of securing messages through cryptography has a long history that may be divided to four main phases:

- From ancient civilization till the beginning of the twentieth century with simple algorithms designed and implemented by hand.
- Around the second world war with the extensive use of electro-mechanical machines.
- In the last fifty years with the widespread use of the computers supported by a mathematical framework.
- The new era of cryptography based on quantum mechanics instead of the use mathematical techniques.

1.1.1 Short History

Ancient Cryptography

The art of cryptography began around 1900 B.C when an Egyptian scribe used a non-hieroglyph for inscription. The earliest known text containing components of cryptography originated in the Egyptian town Menet Khufu on the tomb of Khnumhotep II. The scribe used uncommon hieroglyphic symbols here and there instead of the more commonly used ones. The inscription was not meant to be secret, the transformation was made to dignify it. Only those privileged with an extensive education were able to read and write hieroglyphs. This is the oldest known text to incorporate transformations in the original text though these transformations did not protect the privacy of the text but merely glorified it [2].

In 1500 B.C ancient Assyrian merchants used intaglio, a flat stone with a collage of images and some writing to identify themselves in trading transactions. This mechanism is nowadays known as "digital signature". A particular engraving belonged to a certain trader who was the sole owner of the intaglio that can produce the signature.

During 500–600 B.C., Hebrew transcribers writing down the book of Jeremiah made use of a reversed-alphabet substitution cipher known as ATBASH. The name derives from the first, last, second, and second to last Hebrew letters (Aleph-Tav-Beth-Shin). It works by substituting the first letter of an alphabet for the last letter, the second letter for the second to last, and so on, effectively reversing the alphabet. The ATBASH cipher of the Latin alphabet is given in the Table 1.1 [3]:

Table 1.1 ATBASH cipher of the Latin alphabet

Plaintext	ABCDEFGHIJKLMNOPQRSTUVWXYZ
Ciphertext	ZYXWVUTSRQPONMLKJIHGFEDCBA

In 487 B.C., the Greeks used a device named "skytale" to hide messages. A skytale is a tool used in order to perform a transposition cipher constituting of a cylinder with a thin strip of leather wrapped around it and written on. Once the encryption process is done the leather is taken off and worn as a belt. At the destination, the receiver is assumed to have a matching cylinder. The receiver deciphers the message by wrapping the strip of leather around the cylinder [4].

Around 100–44 B.C., Julius Caesar used a simple substitution method to transform communication with his generals. It was based on three position shift, that is, mathematically, [5]:

$$Y = (X + 3) \bmod 26 \tag{1.1}$$

In Eq. (1.1), X is the alphabet number X (e.g., $X = 1$ for A, $X = 2$ for B, etc.,) and Y is the transformed alphabet. The letter A in the plaintext will thus map into D, and Z into C. This cipher is considered less strong than ATBASH, but it was introduced in a day when few people knew how to read in the first place, so it was strong enough to hide the content of the message.

Around 725–790 A.D, Abu Abu `Abd al-Rahman al-Khalil ibn Ahmad ibn `Amr ibn Tammam al Farahidi al-Zadi al Yahmadi authored a currently lost book on cryptology. His book was inspired by his solution of a cryptogram (i.e., an encrypted message) in Greek for the Byzantine emperor. His solution used what is currently known as the known plaintext attack; this same cryptanalytic method was used in World War II against Enigma messages [6].

In 1379, Gabriel di Lavinde compiled a combination of substitution alphabet and small code at the request of Clement VII. Di Lavinde's collection of Vatican ciphers were, at heart, monoalphabetic ciphers, many also included "nulls", which are special cipher shapes that code for nothing at all, and were added into cipher texts specifically to try to misdirect cryptanalysts [7].

In addition, many of the ciphers in Gabrieli de Lavinde's cipher register also contained a nomenclator; this was typically a list of a dozen-or-so shapes enciphering entire words, like a cross between a cipher and a code. However, it is not clear whether nomenclators were added in the 14th century for security, speed or brevity [8].

In 1466, Leon Battusta Alberti invented the first polyalphabetic cipher. The Alberti cipher was described in Alerti's treatise De Cifris. It uses a device called Formula (known to us as the Captain Midnight Decoder Badge) made up of two concentric disks, attached by a common pin, which can rotate one with respect to the other. The larger disc is called Stabilis and the smaller one is called Mobilis. This class of cipher was not broken until the 1800s [9].

In 1553, Giovan Batista Bellaso La Cifra De Sig, in his publication, described a text autokey cipher that was considered unbreakable for four centuries. He created a new technique of using the Tabula Recta in combination with a passphrase distinct from the encoded message. At the time, it proved to be nearly uncrackable, so he published the method to share it with the world [10].

However, Bellaso's book was not that popular until just a little over thirty years later when Blaise de Vigenère presented Bellaso's method to the court of King Henry the 3rd of France as the Autokey Cipher. Due to its immunity to cryptanalytic attacks, the code became an overnight success, and was named after Vigenère. In some circles, the Tabula Recta is still known as the Vigenère Square.

In 1563 Giovanni Battista Porta published a text on ciphers where he introduced the digraphic cipher. In addition, Giovanno classified ciphers into three main parts: transposition, substitution, and symbol substitution. He also suggested to mislead a cryptanalyst by using synonyms as well as intentionally misspell the plaintext message [7].

In 1586, the French diplomat Blaise de Vigenère published his description of a polyalphabetic cipher similar to the Caesar cipher. In the Vigenère cipher, each letter of the alphabet is shifted along some number of places [11]. This consists of several Caesar ciphers in sequence with different shift values where a table of alphabets was used to encipher. Another more modern substitution cipher was introduced in 1926 by Lester S. Hill and called Hill Cipher. The Hill Cipher was the first polygraphic substitution cipher that was practical to operate on more than one symbols at once [5]. This has a major advantage in making the frequency attack much more difficult by masking the frequency distribution of the letters.

In 1623, Sir Francis Bacon introduced the Baconian cipher [12]. The Baconian cipher uses techniques of steganography and substitution. It is a bilateral cipher known today as the 5-bit binary encoding. To encode a message, each letter of the plaintext is replaced by a group of five of the letters 'A' or 'B'. This replacement is a binary encoding and is done according to the alphabet of the Baconian cipher, shown in Table 1.2.

In 1790, Thomas Jefferson devised an ingenious and secure method to encode and decode messages using *the wheel cipher* [13]. Jefferson's wheel cipher consisted of twenty-six cylindrical wooden pieces, each threaded onto an iron spindle. The letters of the alphabet were inscribed on the edge of each wheel in a random order. Turning these wheels, words could be scrambled and unscrambled. The wheel cipher was later reinvented and used By the US Army in World War II under the name of Strip Cipher.

Table 1.2 Baconian cipher 5-bit encoding

a	AAAAA	g	AABBA	m	ABBAA	s	BAABA	y	BBAAA
b	AAAAB	h	AABBB	n	ABBAB	t	BAABB	z	BBAAB
c	AAABA	i	ABAAA	o	ABBBA	u	BABAA		
d	AAABB	j	ABAAB	p	ABBBB	v	BABAB		
e	AABAA	k	ABABA	q	BAAAA	w	BABBA		
f	AABAB	l	ABABB	r	BAAAB	x	BABBB		

War Driven Cryptography—WWI

British cryptographers came across a German encoded telegram for the first time in 1917. The telegram is referred to as Zimmerman Telegram [14]. British cryptanalysts were able to decipher this telegram and change the history of cryptanalysis by doing so. It is believed that with the use of the deciphered message they were able to convince the United States to join the first word war.

The Zimmerman telegram, shown in Fig. 1.1, was a diplomatic communication between the Foreign Secretary of the German Empire, Arthur Zimmerman, and the German ambassador in Mexico, Heinrich von Eckardt. The telegram offered Mexico the chance to reclaim its territory of New Mexico, Texas, and Arizona in case they join the Germans in WWI. Up until that point during WWI, the United States of America had remained neutral despite requests from the British and their allies. After receiving the deciphered telegram on February 24, 1917, the United Stated joined WWI on April 6, 1917.

As the First World War went by, the United States had the continuous problem of lack of security. The Germans could intercept almost every phone call, leaving the allies moves discreetly known to the Germans. Captain Lewis, the army commander, devised a plan to overcome this problem by the use of the American Indian languages. He used eight Choctaw men he found earlier in the battalion to talk to each other over the radio and phone lines. Within 24 h of the use of Choctaw language as encryption, the advantage fell in favor of the United States.

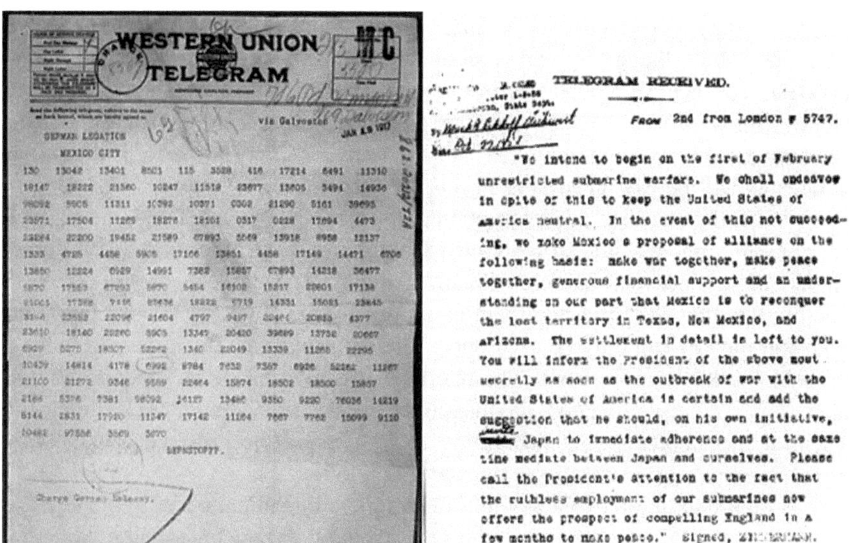
WESTERN UNION TELEGRAM

via Galveston

GERMAN LEGATION

MEXICO CITY

BERNSTORFF.

TELEGRAM RECEIVED.

From 2nd from London # 5747.

"We intend to begin on the first of February unrestricted submarine warfare. We shall endeavor in spite of this to keep the United States of America neutral. In the event of this not succeeding, we make Mexico a proposal of alliance on the following basis: make war together, make peace together, generous financial support and an understanding on our part that Mexico is to reconquer the lost territory in Texas, New Mexico, and Arizona. The settlement in detail is left to you. You will inform the President of the above most secretly as soon as the outbreak of war with the United States of America is certain and add the suggestion that he should, on his own initiative, invite Japan to immediate adherence and at the same time mediate between Japan and ourselves. Please call the President's attention to the fact that the ruthless employment of our submarines now offers the prospect of compelling England in a few months to make peace." Signed, ZIMMERMANN.

Fig. 1.1 Encoded and decoded Zimmerman telegram

War Driven Cryptography—WWII

Arhtur Scherbius invented the Enigma around the end of WWI. The Enigma is an electro-mechanical machine that was used for encryption and decryption of secret messages. The Enigma allowed up to 10^{114} possible configurations. It had several rotors and gears and was virtually unbreakable using brute force methods.

Around 1933–1945, the Enigma was taken and improved by the Nazi Germany. It became their cryptographic workhorse even though it was not considered a commercial success. Later, the Enigma was broken by the polish mathematician Marian Rejewski.

In the meantime, when the allied forces were focused on breaking the enigma machine, the Japanese developed an encryption machine called Purple in 1937. The chief designer of Purple was Kazuo Tanabe and his engineers were Masaji Yamamoto and Eikichi Suzuki. They used stepping switches in contrast to the Enigma machine which used rotors. William Firedman and his team built a replica of Purple based on encrypted messages they recovered. But since no one ever saw a purple machine and no one had an idea how it worked, using it proved to be very difficult. Later on, the team was able to figure out the encryption method used by Purple, and decrypt the encrypted message using a different machine they built. This advancement allowed successful interception of Japanese diplomatic secrets by the United States in WWII.

Modern Encryption

The era of modern cryptography can be divided into two main parts. Part one is the era of symmetric key encryption, where a sender and a receiver use a secretly pre-shared key to establish secure message exchanges. In case of symmetric key encryption, both the sender and receiver use the same key to encrypt and decrypt data. Part two is the era of asymmetric key encryption, where a publicly known key along with a private key are used to establish secure communication transfer. Asymmetric key cryptography can also be used as a way to perform digital signatures as will be explained later in this section.

In 1900, the one-time pad encryption algorithm was invented. The one-time pad encryption is unbreakable. It is derived from a previous cipher called the Vernam cipher, named after its inventor Gilbert Vernam. The unbreakable aspect of the one-time pad comes from two main assumptions: the key used is completely random and the key can only be used once. The security of the one-time pad relies on keeping the key totally secret. The one-time pad uses the XOR modular addition operation. At the sending end, the message is first combined with the key elements. Then, at the receiving end, decryption is done using the same key as shown in Fig. 1.2.

It is important to note that any non-randomness that might occur in the key used in a one-time pad cipher decreases the security and thus the cipher will no longer be considered unbreakable.

The area of modern cryptography really begins with Claude Shannon, with the publication of his paper in 1949. The paper titled "*Communication Theory of*

Fig. 1.2 Example of one-time pad operation

Secrecy Systems" was later followed by the book *Mathematical Theory of Communication*, with Warren Weaver [15]. Claude Shannon established a solid theoretical basis for cryptography and cryptanalysis. Confusion and Diffusion are the two important principles governing his theory [5]. The goal of confusion is to complicate the relation between the key and the cipher text as much as possible, whereas diffusion spreads the influence of one single plaintext bit over multiple cipher text bits [16].

In March 1975, the first draft of DES (Digital Encryption Standard), which is a form of symmetric cryptography, was published in the U.S. Federal Register. DES was proposed by IBM to develop secure electronic communication facilities for businesses. In DES, data are encrypted in 64-bit blocks (shown in Fig. 1.3) using a 56-bit key. The DES algorithm transforms a 64-bit input binary sequence into a 64-bit output sequence. In order to decrypt the message, the same key is used with the same steps in reverse order.

In 2001, the AES (Advanced Encryption System) was published by the National Institute of Standards and Technology (NIST). It is a symmetric block cipher intended to replace DES. All AES operations are performed on 8-bit bytes. The arithmetic operations of addition, multiplication, and division are performed on the finite field GF (2^8). The cipher takes a plaintext block size of 128 bits. The key used can be 128, 192, or 256 bits long. The algorithm used is referred to as AES-128, AES-192, or AES-256. AES encryption and decryption are shown in Fig. 1.4.

In the mid-1970s, a major advance in cryptography occurred with the invention of public-key cryptography. In 1976, the paper titled *New Directions in Cryptography* by Whitfield Diffie and Martin Hellman introduced a radically new scheme for distributing cryptographic keys, and became known as Diffie-Hellman key exchange [17]. In addition, the authors also put forward the idea of authentication by means of a one-way function.

Based on the work of Diffie and Hellman, a new public key encryption algorithm was introduced. This algorithm is known as RSA (shown in Fig. 1.5). It was named

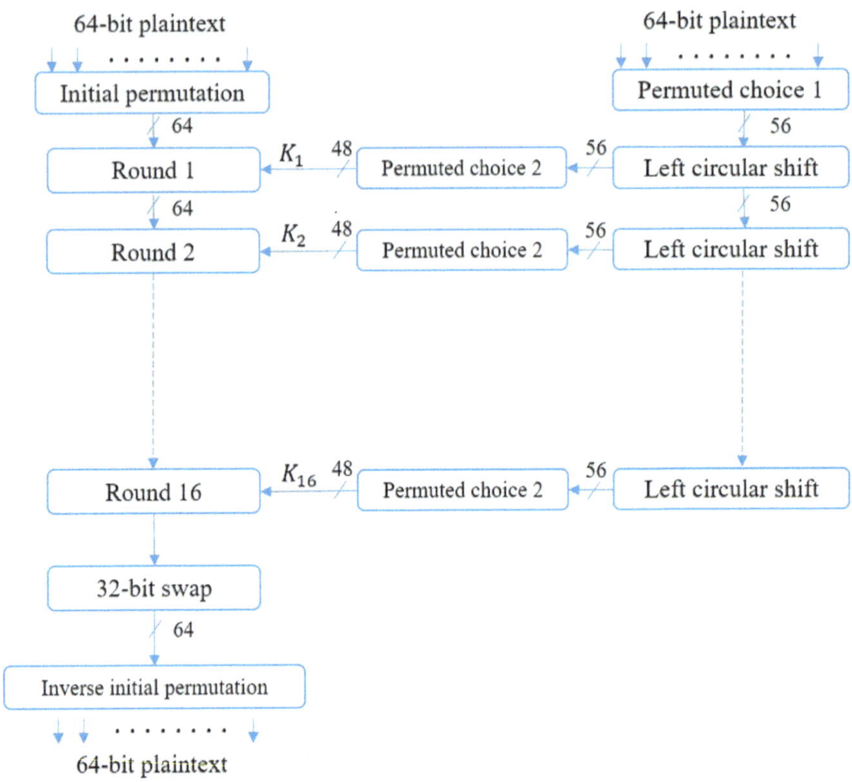

Fig. 1.3 General depiction of DES encryption algorithm

after the three inventors, Ron Rivest, Adi Shamir and Leonard Adelman. The security of RSA is based on the mathematical difficulty of factoring large numbers into their prime components, a major computational task way beyond the capacity of the then existing computers and algorithms [5, 18].

The RSA is a practical public key cipher for both confidentiality and digital signatures, based on the difficulty of factoring large numbers. The steps of operation of RSA are shown in Fig. 1.5.

In 1984, T. El Gamal introduced a public key scheme based on the Diffie-Hellman technique. This cryptosystem can also be used for digital signatures. The operation of the El Gamal algorithm is depicted in Fig. 1.6.

In 1990, Xuejia Lai and James Massey Published “A Proposal for a New Block Encryption Standard”, where they proposed the International Data Encryption Algorithm (IDEA) as a replacement for DES [19]. IDEA functions on 64-bit blocks using a 128-bit key, and comprises of a sequences of eight similar transformations and an output transformation. The processes for encryption and decryption are similar. IDEA bases the majority of its security on interleaving operations from

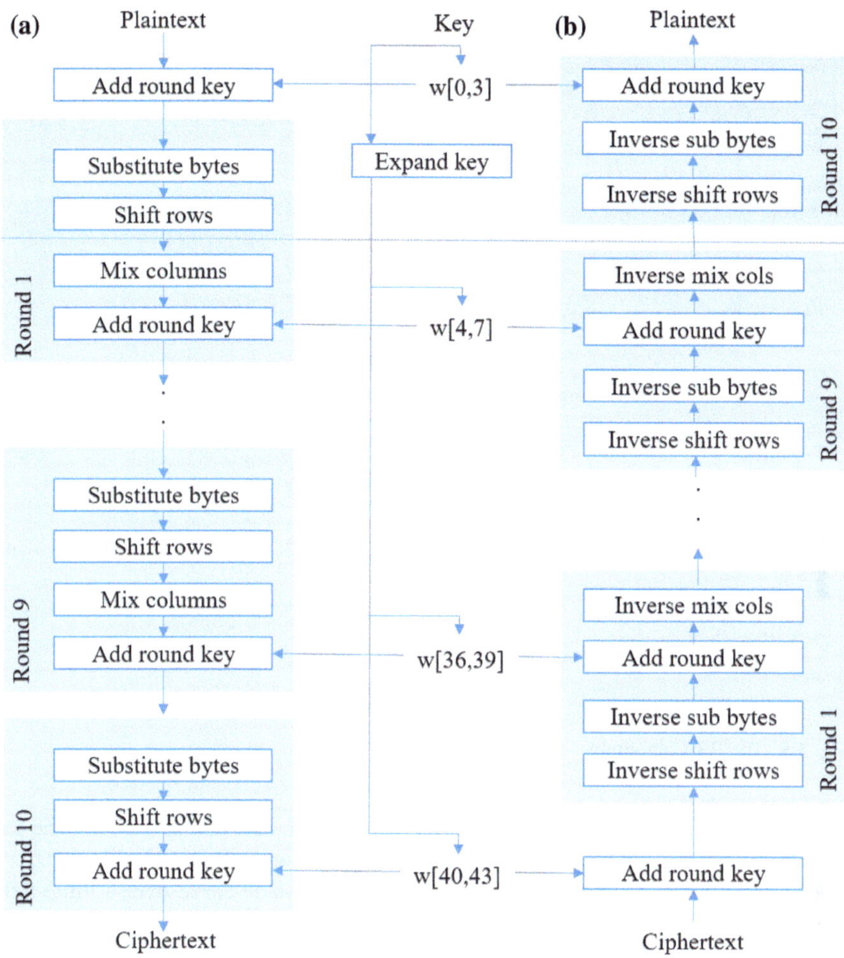

Fig. 1.4 AES. **a** Encryption and **b** decryption

dissimilar groups—modular addition and multiplication, and bitwise eXclusive OR (XOR)—which are algebraically "incompatible" [19].

In 1991, Phil Zimmermann released his first version of Pretty Good Privacy (PGP) [20]. PGP is an encryption protocol that offers cryptographic privacy and authentication for data communication. PGP is often used for signing, encryption and decryption of texts, files, directories, etc., as well as to increase the security of e-mail communication. PGP encryption uses a sequential combination of hashing, data compression, symmetric-key cryptography, and finally public-key cryptography; each step using one of several supported algorithms. Each public key is bound to a user name and/or an e-mail address. The first version of this protocol was released as a response to the threat by the FBI to demand access to communications of the public. PGP was released as a freeware and became a worldwide standard.

Key Generation

Select p, q	p and q both prime, $p \neq q$
Calculate $n = p \times q$	
Calculate $\phi(n) = (p-1)(q-1)$	
Select integer e	$\gcd(\phi(n), e) = 1; 1 < e < \phi(n)$
Calculate d	$d = e^{-1}(mod\phi(n))$
Public key	$PU = \{e, n\}$
Private key	$PR = \{d, n\}$

Encryption

Plaintext:	$M < n$
Ciphertext:	$C = M^e mod\ n$

Decryption

Ciphertext:	C
Plaintext:	$M = C^d mod\ n$

Fig. 1.5 The RSA algorithm

Elliptic curve cryptography algorithms entered wide use in 2004–2005. The use of elliptic curves in cryptography was suggested independently by Neal Koblitz and Victor S. Miller in 1985 [21]. An elliptic curve is a plane curve over a finite field (rather than the real numbers) which consists of the points satisfying Eq. (1.2),

$$y^2 + axy + by = x^3 + cx^2 + dx + e \tag{1.2}$$

Several discrete logarithm-based protocols have been adapted to elliptic curves [22] such as the elliptic curve Diffie–Hellman (ECDH) key agreement scheme based on the Diffie–Hellman scheme, the Elliptic Curve Integrated Encryption Scheme (ECIES), also known as Elliptic Curve Augmented Encryption Scheme or simply the Elliptic Curve Encryption Scheme, the Elliptic Curve Digital Signature Algorithm (ECDSA) based on the Digital Signature Algorithm, the deformation scheme using Harrison's p-adic Manhattan metric, the Edwards-curve Digital Signature Algorithm (EdDSA) based on Schnorr signature which uses twisted Edwards curves, the ECMQV key agreement scheme based on the MQV key agreement scheme, and the ECQV implicit certificate scheme. Figure 1.7 depicts the ECDH key exchange algorithm.

Global Public Elements	
q	prime number
α	$\alpha < q$ and α a primitive root of q

Key Generation by Alice	
Select private X_A	$X_A < q - 1$
Calculate Y_A	$Y_A = \alpha^{X_A} mod\ q$
Public key	$PU = \{q, \alpha, Y_A\}$
Private key	X_A

Encryption by Bob with Alice's Public key	
Plaintext:	$M < q$
Select random integer k	$k < q$
Calculate K	$K = (Y_A)^k mod\ q$
Calculate C_1	$C_1 = \alpha^k mod\ q$
Calculate C_2	$C_2 = KM mod\ q$
Ciphertext	(C_1, C_2)

Decryption by Alice with Alice's Private key	
Ciphertext:	(C_1, C_2)
Calculate K	$K = (C_1)^{X_A} mod\ q$
Plaintext:	$M = (C_2 K^{-1})\ mod\ q$

Fig. 1.6 El Gamal public key scheme

For more information about the protocols depicted in Figs. 1.2, 1.3, 1.4, 1.5, 1.6 and 1.7, interested readers are referred to [5].

1.1.2 Classical Cryptography Limitations

Symmetric Key Encryption Limitations

Throughout history there has been one central problem limiting the widespread use of cryptography. That problem is key management [23]. In cryptographic systems, the term *key* refers to a numerical value used by an algorithm to alter the plaintext,

Global Public Elements

$E_q(a, b)$	elliptic curve with parmeters a, b *and* q, where q is a prime or an integer of the form 2^m
G	point on elliptic curve whose order is large value n

User A key Generation

Select private n_A	$n_A < n$
Calculate public P_A	$P_A = n_A \times G$

User B key Generation

Select private n_B	$n_B < n$
Calculate public P_B	$P_B = n_B \times G$

Calculation of Secret Key by User A

$K = n_A \times P_B$

Calculation of Secret Key by User B

$K = n_B \times P_A$

Fig. 1.7 ECC Diffie-Hellman key exchange same comments as before

making that plaintext information secure by making it understandable only to the individuals who have the corresponding key to recover the information. Consequently, the term key management refers to the secure administration of keys to provide them to users where and when they are required. Historically, encryption systems used what is known as symmetric cryptography. Using symmetric cryptography, it is safe to send encrypted messages without fear of interception (because an interceptor is unlikely to be able to decipher the message); however, there always remains the difficult problem of how to securely transfer the key to the recipients of a message so that they can decrypt the message. Furthermore, in symmetric cryptography, as the number of users increases on a network, the number of keys required to provide secure communications among those users increases rapidly [5]. The invention of public-key cryptography was of central importance to the field of cryptography and provided answers to many key management problems for large scale networks.

Asymmetric Key Encryption Limitations

As discussed before, one major advancement in the field of cryptography is the invention of public-key cryptosystems; the primary feature of public-key cryptography is that it removes the need to use the same key for encryption and decryption

[5]. With public-key cryptography, keys come in pairs of matched "public" and "private" keys. The public portion of the key pair can be distributed in a public manner without compromising the private portion, which must be kept secret by its owner. An operation (for example, encryption) done with the public key can only be undone with the corresponding private key. Prior to the invention of public-key cryptography, it was essentially impossible to provide key management for large-scale networks. For all its benefits, however, public-key cryptography did not provide a comprehensive solution to the key management problem.

Public key cryptography is based on the use of one way functions [24]. One-way functions are mathematical functions that are easy to compute in one direction but difficult to reverse. The ease of calculating the inverse of a function is determined by the time it takes to accomplish this task. If the time increases exponentially with the size (the number of bits) of the input, the one-way function is deemed to be strong. It is worth noting that the perceived security of these ciphers is based on the following [25]:

(1) They have withstood the test of time since there is no published algorithm that will provide a shortcut to breaking the cipher.
(2) In the face of continuous increase in computing power, a cipher declared secure today might not be secure in the future computing environment, which implies that continuous improvement of the algorithm is necessary and the algorithm must allow for this.

Popularly used asymmetric key encryption algorithms include El Gamal, RSA, and Rabin [26, 27].

1.1.3 Quantum Cryptography as a Solution

The shortcomings of classical cryptography can be stated as follows:

(1) The security of a one-way function is not mathematically proven. Future developments in mathematics might result in more efficient algorithms that would invert the one-way function. In addition, cryptography based on one-way functions are vulnerable to increase in computing power which makes brute-force attacks more feasible.
(2) The security of the key during key generation and transmission cannot be absolutely guaranteed. This is true regardless of whether the key is chosen at random from a suitably chosen set of possible values, or generated using a function.
(3) Once a cipher has been compromised, there is no obvious method by which the participants in the secure communication can determine that a breach has occurred [28].

Due to the facts stated above, researchers started looking for better security mechanisms. In the meantime, cryptanalysts continued their search for better cryptanalytic tools and algorithms. A quantum computer has the potential to search the key space in a short period compromising the security of conventional cryptographic systems. As one example, while it takes the classical computer $O(N)$ time to solve the unsorted database search problem, with Grover's quantum database search algorithm, it only takes $O(\sqrt{N})$ time, where N is the number of the entries in the unsorted database [29].

Quantum Cryptography was first proposed in 1970 in a paper that remained unpublished until 1983 [30]. The first protocol in quantum cryptography was proposed in a paper published in 1984 by Bennett and Brassard. Essentially, quantum cryptography was proposed to solve the problem of secure key distribution. Therefore, it is generally known as quantum key distribution (QKD) or quantum key exchange [23] system.

The basics of the quantum key distribution rest on two main principles: the Heisenberg uncertainty principle and the no-cloning theorem [16, 24].

1.2 Quantum Cryptography

Most of our modern systems rely on the usage of symmetric and asymmetric cryptography in tandem. Public cryptography is used as the means for key distribution, while the distributed keys are used in a symmetric encryption system to encode/decode messages. Thus, the security of these systems hinges on the existence of strong one-way functions. The strength of one-way functions proposed to date has not been formally proven. More specifically, the possibility of quantum computers using an efficient factorization algorithm, in the case of asymmetric encryption schemes, cannot be ruled out. Thus, the cryptographic techniques in use today are vulnerable. Furthermore, the cryptanalysis techniques that emerge can be applied retroactively, risking all applications of the past.

Quantum physics has changed the landscape of cryptography in the last two decades. This has been done by introducing quantum cryptography; in other words, cryptosystems that are based on quantum physic to provide unconditionally secure data transfer. Quantum Cryptography (QC) is mainly used for key distribution and called Quantum Key Distribution (QKD). QC and QKD are discussed in the next section.

1.3 Quantum World

Anyone who is not shocked by quantum theory has not understood it.

—Niels Bohr

Quantum information theory was established in the beginning of the last century. Its fundamental concept relies on quantum bits, or qubits, for short. Classical physics cannot be used to describe the behavior of a quantum system. The unique and counter-intuitive aspects of quantum theory are exploited to the benefit of quantum cryptography. Quantum cryptography makes use of the unique properties of qubits to securely transfer messages.

The fundamental concepts that quantum cryptography relies on are as follows:

Information gain versus disturbance: This aspect of quantum physics forms "the engine that powers quantum cryptography" [31]. In quantum cryptography, and in the context of this book, the bits of a message are encoded using photon polarization. To gain any information about the polarization of these qubits, an intruder must observe them. In other words, an intruder should measure the state of the communicated qubit. Such a measurement is destructive and disturbs the qubit's state irretrievably. Consequently, an eavesdropper tapping on the quantum channel can be noticed in a statistically detectable way. This fundamental concept does not apply to classical cryptography where a classical bit can be read without disturbing it.

An unknown quantum state cannot be copied: This fact is formalized in the no-cloning theorem [32]. The no-cloning theorem states that the unknown state of a photon cannot be copied exactly and deterministically. It is worth noting that this fact constitutes a major security feature in quantum cryptography. It is a distinguishing characteristic not applicable to the classical information domain.

1.3.1 Polarization Concept

Photons are the most popular carrier of quantum bits; they have an intrinsic property called polarization [33]. The concept of polarization and its importance in quantum cryptography will be described briefly.

Light is an electromagnetic wave that propagates in a medium. This electromagnetic wave is composed of photons, and it can be described by an electric field and a magnetic field that are perpendicular and orthogonal to each other. Light waves exhibit the phenomenon of polarization. When we consider the polarization state of light, we only need to consider one of its components, either the magnetic or the electric field, since they are correlated and the knowledge of one is equivalent to knowledge of the other. Usually the electric field is considered when talking about a polarization state.

Light may be either polarized or un-polarized. According to the projection of the electric field vector on the plane perpendicular to the travel direction of the light, polarized light can be either linearly polarized, circularly polarized or anywhere in between when it's elliptically polarized, as shown in Fig. 1.8.

Choosing two linear orthogonal polarization axes, we can represent a vertically polarized photon by $|V\rangle$, and a horizontally polarized photon by $|H\rangle$. The general state of polarization is described by the Dirac notation, discussed more fully in Chap. 2. This state is represented as,

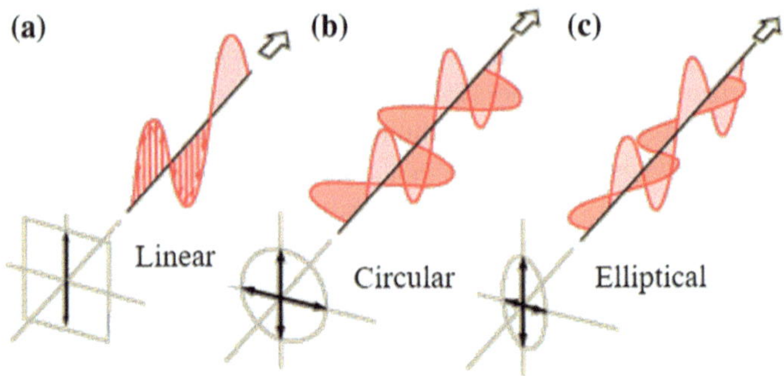

Fig. 1.8 **a** Linear, **b** circular and **c** elliptical polarizations of light

$$|A\rangle = \alpha_V|V\rangle + \alpha_H|H\rangle \tag{1.3}$$

where α_V and α_H are the probability amplitudes and $|\alpha_V|^2 + |\alpha_H|^2 = 1$ [34]. Thus, the state of the photon polarization is mathematically described by a vector of a unit length.

Any two orthogonal polarizations form a basis and the photon polarization can be expressed in terms of that basis. For example [34], linearly polarized photons along the diagonals can be written as:

$$\left|\frac{\pi}{4}\right\rangle = \frac{(|H\rangle + |V\rangle)}{\sqrt{2}}$$

$$\left|-\frac{\pi}{4}\right\rangle = \frac{(|H\rangle - |V\rangle)}{\sqrt{2}}$$

In conventional quantum cryptography, we mainly use two bases: the (+) or rectilinear basis which is $\{|H\rangle, |V\rangle\}$ basis, and the ($\times$) or diagonal basis which is $\{|\frac{\pi}{4}\rangle, |-\frac{\pi}{4}\rangle\}$ basis.

Photons, individually, are completely polarized; their polarization state can be linear, circular or elliptical. Suppose we want to prepare a stream of horizontally polarized photons by sending them through a horizontal polarizing filter. Furthermore, suppose that we want subsequently to measure the polarization of those photons by sending them through a second polarizing filter. We find that only when the measurement filter is in vertical position, no photons pass through it. For all the other orientations, some photons will pass through it.

According to quantum physics laws, each photon in a stream has a certain probability to pass through the measurement filter. The probability is dependent on the orientation of the measurement filter and varies from 0 to 1. Taking the same example as above when the light is horizontally polarized, each photon will have a probability 1 to pass through the measurement filter when it is in the horizontal position. This probability will decrease to $\frac{1}{2}$ at $\frac{\pi}{4}$ and to 0 when the filter is vertically positioned [35].

1.3.2 Quantum Cryptography

Quantum cryptography or quantum key distribution exploits laws of physics described earlier to guarantee unconditionally secure information transfer from a sender, Alice, to a receiver, Bob. It enables Alice and Bob to share a random secret key that can be used for encryption and authentication functionalities. QKD promises unconditional security unlike its classical counterpart that relies on computational complexity. The first implementation of QKD was in the form of BB84 [36]. As mentioned earlier, BB84 was proposed by Bennet and Brassard in 1984. They were the first to realize that quantum states can best exchange information rather than store it. The latter was first proposed by Stephen Wiesner in 1970. Wiesner proposed the idea of Quantum Money by having bank notes which would be impossible to forge. This would require assigning a series of isolated two-state quantum markers to each bank note in addition to its unique serial number [37]. A hypothetical example is attaching to the bank note photons in one of four polarizations states: 0°, 45°, 90° and 135°. Each of these is a two-state system in one of two bases: the rectilinear basis has states with polarizations at 0° and 90° to the vertical, and the diagonal basis has states at 45° and 135° to the vertical.

At the bank, a record of all the polarizations and the corresponding serial numbers would be present. The serial number is printed on the bank notes, while the polarizations states are kept secret. This means that the bank can always verify the polarizations without introducing any disturbance, whereas the counterfeiter cannot. The system proposed by Wiesner is not practical, however, but the underlying thought led to the discovery of several Quantum Key Distribution protocols that are in use today.

In addition to the BB84, several other protocols were proposed as alternatives, such as B92 [38], the six-state protocol [39], BBM92 [40], and SARG04 [41]. QKD needs both a quantum channel as well as a classical channel. The communicating parties do need to have a previously established means of authenticating each other. Unconditionally secure classical authentication schemes do exist and an example is the Wegman-Carter authentication scheme [42, 43]. QKD and other quantum mechanics based protocols are discussed in Chaps. 2 and 3. Here we simply note that QKD provides a solution to a problem not solvable using classical means.

Need for QKD Based Crypto-systems

One of the most important issues facing QKD is whether it is needed at all! Researchers have a range of different opinions. Some of them consider QKD as a solution looking for a problem with no existing or forthcoming use [44–46]; the rest regard QKD as the rescuer because cryptography as we know it is destined to fail [47].

Cryptography is the success story of the information security world. If properly implemented, it can enable sensitive information to be transmitted securely in an insecure environment. For practical purposes, security as known today can prove to be an extremely strong defense mechanism. A system failure might be due to poor key management or human failure rather than due to a cryptographic scheme failing.

The promise of unconditional security through QKD can be ideal for certain applications e.g., banking. However, for it to have a widespread business appeal, it must be directed into solving a business problem, save money and/or make a procedure more efficient. Research and development continue to increase the efficiency of QKD algorithms as well as reduce the cost of the equipment needed for deployment of such technology.

Bruce Schneier states that "Security is a chain: it is as strong as its weakest link" [44]. This is why he described QKD "as awesome as it is pointless" [44]. Strengthening the strongest link will prompt hackers to look somewhere else in the network for vulnerabilities. With this in mind, a QKD proponent's claims should be analyzed carefully. The events that will influence the adoption of QKD are summarized as follows:

(1) Increasing number of applications where currently used cryptographic techniques are considered ineffective and not secure enough.
(2) Advances in mathematical techniques that constitute a threat to currently used cryptographic techniques. One example is the speed with which a large composite number can be factored into its prime components.
(3) Availability of an appropriately functional quantum computer.

Cryptographic techniques can, possibly, never be unconditionally secure, especially when we consider implementation with real devices; this is the case with QKD as well. But QKD may be the way to provide the highest level of security.

Unconditional Security and its Conditions

The main reason behind the need of QKD is the fact that it can achieve unconditional security. In other words, the security of a QKD protocol can be proved without imposing any constraint on the eavesdropping techniques available to Eve. The eavesdropper must interact with the quantum system in order to gain information about the transmitted state. Such interactions can be quantitatively measured. For instance, when Alice uses randomly chosen non-orthogonal states to encode the transmitted message, Eve's intervention will certainly modify the encoded state resulting in errors observable at both Alice's and Bob's ends. Such errors place a limit on the information that can be gained by Eve.

At this point one should mention that the term unconditional security is different than absolute security; security in an absolute sense does not exist. In reality, the unconditional security claimed in QKD exists only under certain conditions. The requirements for unconditionally secure QKD are as follows [48]:

1. An eavesdropper, Eve, cannot intrude Alice's and Bob's devices. In addition, he/she cannot tamper with their setting choices, such as the basis choice.
2. The random number generator must be fully trusted by Alice and Bob. This generator is used to select the states to be sent by Alice and the measurement basis choice by Bob.
3. Unconditionally secure authentication protocols [49] must be used to authenticate the classical channel.
4. An eavesdropper has to obey the laws of quantum physics. In other words, the security of the QKD protocols is based on a restricted set of quantum physics laws.

The failure of these requirements would compromise the security of a QKD protocol. However, it must be noted that even if all the conditions stated above are met, unconditional security is not guaranteed at the implementation level. In addition, the implementation must be free of any unwanted information leakage.

Limitations of Quantum Key Distribution

The notion of key distribution using quantum states is a significant development in the field of security. It is the only way of securing communication in the era of quantum computers. However, the technology that can realize QKD is still in its immature phase and has several restrictions. The limitations associated with a system implementing the BB84 QKD protocol can be described as follows:

1. Photon Sources:

The security of current implementations of QKD using BB84 is depends on single photon states. In other words, it is essential that Alice generate states of single photon. Otherwise, in cases where multiple photons with the same state are generated, an eavesdropper will be able to launch a photon number splitting attack (PNS). During a PNS attack, the eavesdropper will have access to the additional photon/s generated by Alice and will analyze them and get the information he/she needs. Such an attack will go undetected.

Current implementations use faint laser pulses since it turns out that generating single photons with guaranteed periodicity is difficult. Therefore, as a practical compromise, faint laser sources generate photons such that most time slots will be empty, a few would have single photons, and very few more than a single photon.

2. Distance of communication:

Due to detector noise and fiber losses, the range of current quantum key distribution systems is limited to 60–100 km [50–52]. This limitation is associated with the fact that BB84 and its variants are single photon based protocols with average number of photons per pulse much less than 1.

3. Data rates

In today's fiber optic communication systems, transmission rates on the order of several Gigabits are easily attainable, but that is not the case with QKD. This is due to the limitation of no more than a single photon per pulse causing a large number of empty pulses, as well as the sifting process that eliminates half of the possible key. QKD in conjunction with the one-time pad protocol is thus exploited only for the transmission of most confidential data. However, implementing QKD with AES or any other symmetric cipher is possible and can offer great improvement to the security of any system.

4. Security:

Despite the fact that QKD's unconditional security has been proven, any implementation will be subject to attacks at the device level. Nevertheless, a security breach caused by a flaw in a device could possibly be easier to deal with as device technology matures compared to cryptography based on unproven mathematical assumptions.

1.4 Post-quantum Cryptography

Post-quantum cryptography refers to classical cryptographic algorithms believed to be secure against an attack by a quantum computer once one is implemented. Most popular public-key algorithms cannot withstand such an attack. Popular public-key algorithms can potentially be rendered ineffective by an adequately powerful quantum computer. Popular algorithms being used for secure communication currently rely on one of three mathematically complex one-way functions: the integer factorization problem, the discrete logarithm problem or the elliptic-curve discrete logarithm problem. A sufficiently potent quantum computer is expected to break all the three algorithms.

Currently, more attention is being directed to designing innovative classical cryptography schemes that can withstand the threats of quantum computers, in preparation for the next era of cryptography. The majority of currently used symmetric cryptography algorithms and hash functions are considered to be *relatively* secure against attacks by quantum computers [53]. However, the threat quantum computing poses is on currently used public key based algorithms. Grover's algorithm speeds up quantum attacks against symmetric ciphers, however, doubling the key size can successfully block these attacks [54]. The rest of this section provides examples of different types of classical cryptography that might be able to withstand quantum computer attacks.

1.4.1 Lattice-Based Cryptography

A class of classical cryptography that holds a great promise to be used in the post-quantum cryptography era is Lattice-based cryptography [55]. Such

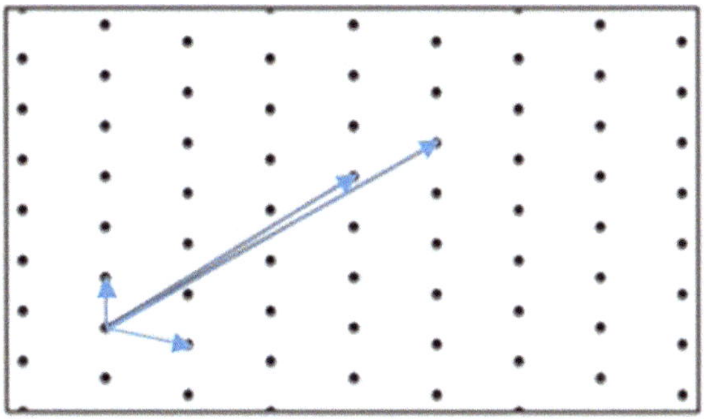

Fig. 1.9 A two-dimensional lattice and two possible bases

cryptography schemes are considered to be simple and are supported by robust security proofs based on worst-case hardness and fairly efficient implementations. Lattice-based cryptography is thought to be able to withstand attacks from quantum computers. Lattice-based cryptography is well studied in [56–60].

A lattice is a set of points in n-dimensional space with a periodic structure, such as the one illustrated in Fig. 1.9.

More formally, given n-linearly independent vectors $b_1, \ldots, b_n \in R_n$, the lattice generated by them is the set of vectors $L(b_1, \ldots, b_n) = (\sum x_i b_i | x_i \in \mathrm{Z})$. The vectors $b_1, \ldots, b_n$ are known as a basis of the lattice.

The application of lattices in cryptography was first discovered in a revolutionary paper by Ajtai [61]. The outcome of his work is now developed into a complete area of research with a focus on expanding the scope of lattice-based cryptography and on generating more useful lattice-based cryptosystems.

The example of lattice-based cryptosystems that attracted the most attention is the "NTRU" public-key-encryption system (1998). NTRU is a ring-based cryptosystem proposed by Hoffstein et al. [62].

1.4.2 *Multivariate Cryptography*

Multivariate cryptography is the class of cryptosystems with public keys constituting of a set of multivariate polynomials. In other words, multivariate Public-key cryptography is the field of PKCs where the trapdoor one-way function is a multivariate quadratic polynomial map over a finite field. The public key generally given by a set of quadratic polynomials as follows:

$P(x_1, \ldots, x_n) = (p_1(x_1, \ldots, x_n), p_2(x_1, \ldots, x_n), \ldots, p_m(x_1, \ldots, x_n))$ and the plaintext is given as $M = (x_1, \ldots, x_n)$. Thus, the cipher text is the following polynomial evaluation: $(M) = (p_1(x_1, \ldots, x_n), p_2(x_1, \ldots, x_n), \ldots, p_m(x_1, \ldots, x_n))$. For the decryption, the receiver needs to know a trapdoor so that it is possible to invert the quadratic map and get the plaintext $(x_1, \ldots, x_n) = P^{-1} = (c_1, \ldots, c_m)$

Among the most well-known multivariate public-key cryptosystems are C^* by Matsumoto and Imai [59]; HFE (Hidden Field Equations) by Patarin [60]; UOV (Unbalanced Oil and Vinegar) by Kipnis et al. [63]; Rainbow/TTS by Ding and Schmidt [64].

1.4.3 Hash-Based Cryptography

Hash-based cryptography is the general term used for schemes of cryptographic primitives using hash functions as a way to secure communication transfer. Up to this point, the use of hash-based cryptography is restricted to digital signatures schemes. Digital signatures provide authenticity, integrity, and non-repudiation of data. They are generally employed in identification and authentication protocols. Hence, the presence of secure digital signature algorithms is critical for conserving IT-security in the post-quantum cryptography era [53].

Digital signatures used today such as RSA [65], DSA [66], ECDSA [67] are not immune to quantum computer threats since they rely on the problem of factoring large integers to provide secure transfer of messages. Similar to any other digital signature scheme, the security of hash-based signature schemes depends on the collision resistance of the hash function in use. The presence of collision resistant hash functions is a condition for the existence of a digital signature scheme that is able sign different documents with the use of a single private key. In other words, a signature scheme maps documents of an arbitrary bit length into strings of fixed bit length called digital signatures. So, digital signature algorithms are in fact hash functions that should be collision resistant. In case there are two documents with the same digital signature, the signature scheme is not considered secure anymore. This indicates that there exist hash-based digital signature schemes as long as there exists any digital signature scheme that can sign different documents with the use of one private key. Thus, one can conclude that hash-based signature schemes are the most significant post-quantum signature candidates. It is important to note that there is no formal proof of hash-function being quantum computer resistant, however, their security requirements are minimal.

Hash-based signature schemes were invented by Ralph Merkle in 1979 [68]. The Merkle's signature scheme is based on hash trees known as Merkle trees. Merkle's scheme is based on the one-time signature scheme of Lamport [69]. The main advantage of the Merkle scheme is the fact that it is robust against the threats imposed from quantum computers using Shorr's algorithm.

1.4.4 Code-Based Cryptography

Code-based cryptography is the set of cryptosystems with an underlying one-way function that uses an error correcting code C. This one-way function may entail adding an error to a message of C or calculating a syndrome according to a parity check matrix of C.

The first example of code-based cryptography is McEliece's hidden-Goppa-code public-key encryption system [70]. This system was first introduced in 1978 by Robert McEliece and is invulnerable to Shor's algorithm. Recent attention is directed to the McEliece cryptosystem by the cryptographic community. That is mainly due to the fact that it is one of the best candidates for post-quantum secure public key cryptography [53]. The McEliece cryptosystem is based on error-correcting linear codes and is described in [71]. McEliece cryptosystem uses binary Goppa codes for messages encryption and decryption. The McEliece cryptosystem has many variants that make use of different linear codes, however, most of them have been proven to be susceptible to cryptanalytic attacks. It is important to note that the original algorithm introduced in 1978 is considered to be secure if adequate parameters are chosen. The main disadvantage associated with the McEliece algorithm is the fact is its key sizes are very large making it inefficient, thus, it has been rarely used in practice. The McEliece PKC is considered as a feasible substitute to RSA and is a secure cryptosystem in the post-quantum world, overall it does merit more analysis and further attention [72].

1.5 Scope and Contributions of This Book

This book is a modest contribution toward fulfilling the insatiable thirst for unconditionally secure information transfer. The earlier sections of this chapter have reviewed the history of cryptography including quantum cryptography or quantum key distribution. We have noted that quantum physics makes it possible for us to share an unconditionally secure *random* information string between two parties. Users' information to be shared between the parties can then be Xor'd with the random key at each end and recover the information. To the extent that the key is random and used *only once,* the information transferred is unconditionally secure.

The problem with this approach is one of practical realizability. Quantum key distribution is effective only over a limited span of distance (a few hundred miles) and a relatively modest rate of key transfer at a few hundred kilobits per second. Both these limitations are incompatible with the market requirements of transferring secure information at higher rates over longer distances.

As mentioned earlier, the current state of the art in quantum key distribution does not allow the transfer of secure keys at a rate equal to the desired information transfer rate. Therefore, quantum key distribution merely refreshes keys at a fast-enough rate so they can be used to secure information via conventional encryption methods. Used in this manner, quantum key distribution does not offer unconditionally secure transfer. It merely increases the burden on cryptanalysts by orders of magnitude.

The BB84 protocol, mentioned earlier in this chapter, is a means to simply detect the presence of an intruder on the communication channel. Once this is assured (i.e., the absence of any pilfering effort is guaranteed), the key is recovered by error correcting techniques to negate the impact of any error on the transmission channel. In one of the later chapters of the book, the authors posit that the presence of an intruder on an optical cable can be detected when one of the channels is carrying data at commercial data transfer rates, say, 10 Gigabits per second. The implication is that such a scheme likely offers a way to offer security at higher data rates and without the encumbrance of conventional encryption and decryption techniques in the process.

The techniques presented in the book are based on photonic rather than quantum techniques. Photonics, based on a multiplicity of photons, is a mature technology. The generation, transfer, and detection of a stream of photons is no longer an art or even science, it's plain engineering. All the techniques presented in this book are based on multiplicity of photons and, hence the name Multi-photon Quantum Secure Communication.

1.6 Organization of This Book

This book explores alternative ways that can accomplish secure information transfer without the need for a quantum channel as in the case of QKD-based techniques. This chapter presents an introduction to the science of cryptography as it has evolved over the past couple of thousand years. Furthermore, it addresses the shortcomings of cryptography as practiced today and points out the need for exploring additional techniques. In particular, it points to the need for quantum mechanics based techniques in cryptography. Chapter 2 of the book provides basic concepts in quantum information science including the notion of a qubit and its quantum states. It presents the characteristics of photons and, in the process, lays the foundation for multi-photon communication.

Chapter 3 of the book discusses the various techniques of Quantum Key Distribution as practiced today. Chapter 4 of the book addresses a class of quantum cryptography called key communication in quantum noise (or KCQ) based on quantum detection and communication theory protocols. Chapter 5 introduces the three-stage protocol. The three-stage protocol is the basic building block of most of the protocols discussed in this book. Chapter 6 generalizes the three-stage protocol into a family of multi-stage protocols. It compares the multi-stage protocol with single-photon protocols and illustrates how a multi-photon protocol can be made secure against man-in-the-middle attack. Since a multi-photon protocol is, in general, subject to photon-siphoning attacks, the protocol introduces another variable to thwart such attacks. Chapter 7 analyzes the security of the multi-stage, multi-photon tolerant protocol for quantum secure communication. The security of the multi-stage protocol is based on the fact that while a legitimate receiver only needs to distinguish between two orthogonal polarization states, an intruder has to distinguish among an indefinite number of possible polarization states. Chapter 8 presents a security analysis of the multi-stage protocol. It analyzes intercept-and-resend and photon number splitting attacks in the multi-stage multi-photon protocol. It lays down the conditions under which the multi-stage multi-photon protocol can approach the strength of a quantum-secure protocol.

Chapter 9 extends the application space of the multi-stage multi-photon protocol to wireless communication. In particular, it examines the viability of using the multi-stage multi-photon protocol for secure key distribution in the IEEE 802.11i protocol.

Chapter 10 discusses an application of the polarization property of light in detecting intrusion on an optical fiber with the objective of stealing information flowing through it. The system discussed in this chapter presents an innovative and cost-effective means to prevent data theft in contemporary telecommunication systems.

Chapter 11 introduces the use of the polarization channel of an optical fiber to transfer data; more specifically, exchange symmetric keys between the two ends of the optical fiber. Use of the symmetric keys will allow any conventional symmetric encryption to take place between any number of data channels supported by the optical fiber. Note that any encryption based on symmetric keys can be only computationally secure but, since the keys can be exchanged at a rapid rate, we can still achieve a high level of security on the data channel.

Chapter 12 presents an ultra-secure router-to-router key exchange system. The key exchange process can be initiated by either router at will and can be carried out as often as required. The cryptographic strength of the proposed protocols lies in the use of multi-stage transmission where the number of variables exceeds the number of stages by one, ensuring that the number of possible measurements is one less that the number of variables. The proposed system carries out all processing in electronics and is not vulnerable to the man in the middle attack.

References

1. Singh, S. (1999). *The code book: The secret history of codes and code-breaking* (1st ed.). Great Britain: Fourth Estate.
2. Damico, T. M. (2009). A brief history of cryptography. *Inquiries Journal/Student Pulse, 1* (11), PG. 1/1.
3. Kile, J. (2013). The Atbash cipher and Jeremiah 51:1. In *Mysterious writings inspiring the search for treasure, mystery, and adventure*, January 20, 2013.
4. Djekic, M. (2013). A scytale—Cryptography of the ancient sparta. *The Best of Australian Science*, November 25, 2013.
5. Stallings, W., & Tahiliani, M. P. (2014). *Cryptography and network security: Principles and practice* (Vol. 6). London: Pearson.
6. Jackob, M. (2001). SANS Info Sec Reading Room. History of Encryption.
7. Mollin, R. A. (2005). *Codes: The guide to secrecy from ancient to modern times*. Boca Raton: CRC Press.
8. Pelling, N. (2016). Fifteenth century cryptography. *Cipher Mysteries*, July 6, 2016.
9. Whitman, M. E., & Mattord, H. J. (2011). *Principles of information security*. Boston: Cengage Learning.

10. Buonafalce, A. (2006). Bellaso's reciprocal ciphers. *Cryptologia, 30*(1), 39–51.
11. Guenther, C. (2003). *The relevance of quantum cryptography in modern cryptographic systems*. GSEC Practical Requirements (v1. 4b). http://www.giac.org/practical/GSEC/Christoph%20Guenther%20GSEC.pdf.
12. Salomon, D. (2003). *Data privacy and security: Encryption and information hiding*. Berlin: Springer Science & Business Media.
13. Hunter, F. (2011). Thomas Jefferson the cryptographer. *Frances Hunter's American Heroes Blog*, September 7, 2011.
14. Wertheim, A. T. (1967). *The Zimmermann telegram*. London.
15. Shannon, C. E. (2001). A mathematical theory of communication. *ACM SIGMOBILE Mobile Computing and Communications Review, 5*(1), 3–55.
16. Kaeo, M. (2003). *Designing network security*. USA: Cisco Press.
17. Diffie, W., & Hellman, M. (1976). New directions in cryptography. *IEEE Transactions on Information Theory, 22*(6), 644–654.
18. Rivest, R. L., Shamir, A., & Adleman, L. (1978). A method for obtaining digital signatures and public-key cryptosystems. *Communications of the ACM, 21*(2), 120–126.
19. Hoffman, N. (2007). A simplified IDEA algorithm. *Cryptologia, 31*(2), 143–151.
20. Zimmermann, P. (1991). *Why i wrote PGP*. Part of the Original.
21. Sutikno, S., Surya, A., & Effendi, R. (1998). An implementation of ElGamal elliptic curves cryptosystems. In *The 1998 IEEE Asia-Pacific Conference on Circuits and Systems, 1998, IEEE APCCAS 1998*.
22. Martínez, V. G., Encinas, L. H., & Ávila, C. S. (2010). A survey of the elliptic curve integrated encryption scheme. *Ratio, 80*(1024), 160–223.
23. Gisin, N., Ribordy, G., Tittel, W., & Zbinden, H. (2002). Quantum cryptography. *Reviews of Modern Physics, 74*(1), 145.
24. Schneier, B. (2007). *Applied cryptography: Protocols, algorithms, and source code in C*. USA: Wiley.
25. Stallings, W. (2005) *Cryptography and network security: Principles and practice*. USA: Prentice Hall.
26. Salkever, A. (2003). *A quantum leap in cryptography*. July 15, 2003. Available from: http://www.businessweek.com/technology/content/jul2003/tc20030715_5818_tc047.htm.
27. idQuantique SA. (2003). *Breakthrough in quantum cryptography—Swiss partnership to release world's first integrated quantum key infrastructure*. December 15, 2003. Available from: http://www.idquantique.com/files/wise-press-engl.pdf.
28. Koashi, M., & Preskill, J. (2003). Secure quantum key distribution with an uncharacterized source. *Physical Review Letters, 90*(5), 057902.
29. Grover, L. K. (1996). A fast quantum mechanical algorithm for database search. In *Proceedings of the Twenty-Eighth Annual ACM Symposium on Theory of Computing*.
30. Wiesner, S. (1983). Conjugate coding. *ACM Sigact News, 15*(1), 78–88.
31. Fuchs, C. A. (1996). *Information gain vs. state disturbance in quantum theory*. arXiv preprint quant-ph/9611010.
32. Wootters, W. K., & Zurek, W. H. (1982). A single quantum cannot be cloned. *Nature, 299* (5886), 802–803.
33. Huard, S. (1997). *Polarization of light*. Germany: Wiley-VCH.
34. Brassard, G., & Salvail, L. (1994). Secret-key reconciliation by public discussion. In *Advances in Cryptology—EUROCRYPT'93*. Berlin: Springer.
35. antenna-theory.com. http://www.antenna-theory.com/basics/polarization.php. Cited September 28, 2015.
36. Bennett, C. H., & Brassard, G. (2014). Quantum cryptography: Public key distribution and coin tossing. *Theoretical Computer Science, 560*, 7–11.
37. Lo, H.-K., Spiller, T., & Popescu, S. (1998). *Introduction to quantum computation and information*. Singapore: World Scientific.
38. Bennett, C. H. (1992). Quantum cryptography using any two nonorthogonal states. *Physical Review Letters, 68*(21), 3121.

39. Bruss, D. (1998). Optimal eavesdropping in quantum cryptography with six states. *Physical Review Letters, 81*(14), 3018.
40. Bennett, C. H., Brassard, G., & Mermin, N. D. (1992). Quantum cryptography without Bell's theorem. *Physical Review Letters, 68*(5), 557.
41. Scarani, V., Acin, A., Ribordy, G., & Gisin, N., et al. (2004). Quantum cryptography protocols robust against photon number splitting attacks for weak laser pulse implementations. *Physical Review Letters, 92*(5), 057901.
42. Carter, J. L., & Wegman, M. N. (1977). Universal classes of hash functions. In *Proceedings of the Ninth Annual ACM Symposium on Theory of Computing*.
43. Wegman, M. N., & Carter, J. L. (1981). New hash functions and their use in authentication and set equality. *Journal of Computer and System Sciences, 22*(3), 265–279.
44. Schneier, B. (2008). *Quantum cryptography: As awesome as it is pointless*. Cited 10/13/2015. Available from: http://archive.wired.com/politics/security/commentary/securitymatters/2008/10/securitymatters_1016.
45. Schneier, B. (2009). *Schneier on security*. USA: Wiley.
46. Paterson, K. G., Piper, F., & Schack, R. (2007). Quantum cryptography: A practical information security perspective. *Nato Security Through Science Series D-Information and Communication Security, 11,* 175.
47. Ghernaouti-Helie, S., Tashi, I., Laenger, T., & Monyk, C. (2009). *SECOQC business white paper*. arXiv preprint arXiv:0904.4073.
48. Scarani, V., Bechmann-Pasquinucci, H., Cerf, N. J., Dušek, M., Lütkenhaus, N., & Peev, M. (2009). The security of practical quantum key distribution. *Reviews of Modern Physics, 81,* 1301.
49. Stinson, D. R. (2005) *Cryptography theory and practice* (3rd ed.). Boca Raton: CRC press.
50. Gisin, N., Ribordy, G., Tittel, W., & Zbinden, H. (2002). Quantum cryptography. *Reviews of Modern Physics, 74,* 145.
51. idQuantique (September 28). http://www.idquantique.com/qkd.html.
52. I. MagiQ Technologies. (September 28). http://www.magiqtech.com.
53. Bernstein, D. J. (2009). Introduction to post-quantum cryptography. In *Post-quantum cryptography* (pp. 1–14). Berlin: Springer.
54. Bernstein, D. J. (2010). Grover vs. mceliece. In *International Workshop on Post-Quantum Cryptography*. Berlin: Springer.
55. Micciancio, D., & Regev, O. (2009). Lattice-based cryptography. In *Post-quantum cryptography* (pp. 147–191). Berlin: Springer.
56. Kumar, R., & Sivakumar, D. (2001). Complexity of SVP–a reader's digest. *SIGACT News, 32* (3), 40–52.
57. Micciancio, D. (2001, Fall). *Lattices in cryptography and cryptanalysis. Lecture Series*. San Diego: University of California.
58. Micciancio, D. (2009). Cryptographic functions from worst-case complexity assumptions. In *The LLL algorithm* (pp. 427–452). Berlin: Springer.
59. Matsumoto, T., & Imai, H. (1988). Public quadratic polynomial-tuples for efficient signature-verification and message-encryption. In *Eurocrypt'88*, vol. 330 (pp. 419–453). LNCS.
60. Regev, O. (2006). Lattice-based cryptography. In *CRYPTO*. Berlin: Springer.
61. Ajtai, M. (1996). Generating hard instances of lattice problems. In *Proceedings of the Twenty-Eighth Annual ACM Symposium on Theory of Computing*.
62. Hoffstein, J., Pipher, J., & H. Silverman, J. (1998). NTRU: A ring-based public key cryptosystem. *Algorithmic Number Theory (ANTS III)*. 10.1007/BFb0054868.
63. Kipnis, A., Patarin, J., & Goubin, L. (1999). Unbalanced oil and vinegar signature schemes. In *Eurocrypt*. Berlin: Springer.
64. Ding, J., & Schmidt, D. (2005). Rainbow, a new multivariable polynomial signature scheme. In *ACNS*. Berlin: Springer.

65. Ding, J., Cabarcas, D., Schmidt, D., Buchmann, J., & Tohaneanu, S. (2008). Mutant Gröbner basis algorithm. In *Proceedings of the 1st International Conference on Symbolic Computation and Cryptography (SCC 2008)*, Beijing, China, LMIB.
66. Braeken, A., Wolf, C., & Preneel, B. (2005). A study of the security of unbalanced oil and vinegar signature schemes. In *CT-RSA*. Berlin: Springer.
67. Chen, J., & Moh, T. (2001). On the Goubin-Courtois attack on TTM. *Cryptology ePrmt Archive*, Vol. 72.
68. Courtois, N., Daum, M., & Felke, P. (2003). On the security of HFE, HFEv-and quartz. In *Public key cryptography*. Berlin: Springer.
69. Lamport, L. (1979). Constructing digital signatures from a one way function. *Technical Report SRI-CSL-98, SRI International Computer Science Laboratory.*
70. Faugere, J.-C. (1999). A new efficient algorithm for computing Gröbner bases (F 4). *Journal of Pure and Applied Algebra, 139*(1), 61–88.
71. McEliece, R. J. (1978). A public-key cryptosystem based on algebraic. *Coding Thv, 4244,* 114–116.
72. Siim, S. (2015). *Study of McEliece cryptosystem.*

Chapter 2
Mathematical Background

This chapter gives the mathematical background necessary for understanding quantum mechanics used in the rest of the book. The abstract concept of a qubit as the quantum extension of a classical bit is first introduced. Characteristics of photons are then covered to lay the foundation for the multi-photon communication. An exposition of the polarization degree of freedom of photons in the multi-photon regime is made.

2.1 Basic Concepts in Quantum Information

In quantum information and quantum cryptography, the fundamental entity is the concept of a qubit (quantum bit). It is a generalization of a classical bit. Before introducing the notion of the qubit, we need to review how to represent a quantum system mathematically.

2.1.1 Quantum State and Qubit

In quantum physics, the state of an isolated quantum system is described by its quantum state. The quantum state provides the probability distribution for each possible measurement outcome of the quantum system. Mathematically, the quantum state is an entity defined in the linear complex space called the Hilbert space [1]. One can intuitively regard the quantum state as a vector whose magnitude and direction convey all the information about the quantum system.

In the so-called Dirac notation [2], a quantum state is denoted by $|\psi\rangle$. For example, suppose the quantum system can exhibit one of two possibilities (e.g., the horizontal or vertical polarization of a photon, the ground and excited states of a two-level atom, etc.) and we denote the two possibilities as '0' and '1'. Then we can

P. K. Verma et al., *Multi-photon Quantum Secure Communication*,
Signals and Communication Technology,
https://doi.org/10.1007/978-981-10-8618-2_2

represent the quantum system by the state $|0\rangle$ if it is found to be in '0'. Likewise, the system can be represented by the state $|1\rangle$ if it is found to be in '1'. The two quantum states $|0\rangle$ and $|1\rangle$ then represent two different possibilities of the state of the quantum system. In quantum information, a qubit is a quantum state with two possibilities $|0\rangle$ and $|1\rangle$ corresponding to the events '0' and '1' which are mutually exclusive, following the notion of a classical bit. Since a quantum state is an object in the linear complex Hilbert space, the natural interpretation of the mutual exclusiveness of the two events for the quantum states is that $|0\rangle$ and $|1\rangle$ are orthogonal vectors. This concept will be made clear in the following by introducing the scalar product (bracket algebra) in the complex Hilbert space. It should be noted that one may use different notations for a qubit depending on the context, e.g., $|H\rangle$ and $|V\rangle$ for the two orthogonal polarizations of a photon, $|g\rangle$ and $|e\rangle$ for the two energy levels of a two-level atom, etc.

The peculiarity of a quantum system in contrast to its classical counterpart is that the state of a quantum system can be a 'superposition' of the possibilities of the measurement outcomes of the system. Mathematically, this is represented by

$$|\psi\rangle = \alpha|0\rangle + \beta|1\rangle, \tag{2.1}$$

where α and β are complex numbers. Since $|0\rangle$ and $|1\rangle$ correspond to the states for two mutually exclusive events, one may also write them as basis vectors using matrix notation:

$$|0\rangle \to \begin{bmatrix} 1 \\ 0 \end{bmatrix}, \quad \text{and} \quad |1\rangle \to \begin{bmatrix} 0 \\ 1 \end{bmatrix}. \tag{2.2}$$

Using the matrix notation, we have

$$|\psi\rangle = \begin{bmatrix} \alpha \\ \beta \end{bmatrix}. \tag{2.3}$$

The coefficients α and β are called the probability amplitudes and they need to satisfy the condition

$$|\alpha|^2 + |\beta|^2 = 1. \tag{2.4}$$

The interpretation is that there is a probability of $|\alpha|^2$ for the quantum state described by $|\psi\rangle$ to be in state $|0\rangle$ and a probability of $|\alpha|^2$ for $|\psi\rangle$ to be in state $|1\rangle$.

The quantum state $|\psi\rangle$ interpreted as a column vector is also called a 'ket'. The corresponding row vector is called a 'bra' $\langle\psi|$, and is related to a ket by the conjugate transposition, also called the Hermitian conjugate or adjoint:

$$\langle\psi| = (|\psi\rangle)^{\dagger}. \tag{2.5}$$

In matrix notation, we then have

$$\langle 0| = [1 \quad 0], \quad \langle 1| = [0 \quad 1], \quad \text{and} \quad \langle\psi| = \alpha^*\langle 0| + \beta^*\langle 1| = [\alpha^* \quad \beta^*], \tag{2.6}$$

where * denotes the complex conjugation. Hence the conjugate transposition is equivalent to complex conjugation + matrix transposition. With the correspondence to column and row matrices, we can perform all linear algebraic operations with the bras and kets. In particular, we have the scalar product between a bra $\langle\psi|$ and a ket $|\phi\rangle$ defined by:

$$\langle\psi|\phi\rangle \equiv ((\langle\psi|)\cdot(|\phi\rangle)). \tag{2.7}$$

It should be noted that, generally, the scalar product results in a complex number and it satisfies the relation $\langle\psi|\phi\rangle = \langle\phi|\psi\rangle^*$. The norm of the state $|\psi\rangle$ is defined to be

$$\||\psi\rangle\| = \sqrt{\langle\psi|\psi\rangle}. \tag{2.8}$$

By convention, basis vectors are taken to have unit norms. Therefore, for the qubit,

$$\langle 0|0\rangle = \langle 1|1\rangle = 1, \quad \langle 0|1\rangle = \langle 1|0\rangle = 0, \tag{2.9}$$

and

$$\langle\psi|\psi\rangle = |\alpha|^2 + |\beta|^2 = 1 \tag{2.10}$$

as required by the normalization condition of the probability amplitudes. Note that $\langle 0|\psi\rangle = \alpha$ and $\langle 1|\psi\rangle = \beta$. The state of a qubit is always normalized unless otherwise mentioned explicitly.

The basis $\{|0\rangle, |1\rangle\}$ is conventionally called the computational basis. Since a linear combination of quantum states is also a quantum state, one can define different bases for a qubit. For example, the basis,

$$\left\{|+\rangle \equiv \frac{|0\rangle + |1\rangle}{\sqrt{2}}, |-\rangle \equiv \frac{|0\rangle - |1\rangle}{\sqrt{2}}\right\} \tag{2.11}$$

is called the Hadamard (or diagonal) basis. In addition, the qubit $|\psi\rangle$ can be written in different bases. In particular, we observe that

$$\begin{aligned} |\psi\rangle &= \alpha|0\rangle + \beta|1\rangle = \langle 0|\psi\rangle|0\rangle + \langle 1|\psi\rangle|1\rangle = |0\rangle\langle 0|\psi\rangle + |1\rangle\langle 1|\psi\rangle \\ &= \big(|0\rangle\langle 0| + |1\rangle\langle 1|\big)|\psi\rangle, \end{aligned} \tag{2.12}$$

where we make use of the relations $\langle 0|\psi\rangle|0\rangle = \alpha|0\rangle = |0\rangle\alpha = |0\rangle\langle 0|\psi\rangle$ by noticing that $\langle 0|\psi\rangle$ is a scalar and $|0\rangle\langle 0|\psi\rangle = (|0\rangle\langle 0|)|\psi\rangle$ by realizing that the bras and kets are simply row and column vectors. The operation $|0\rangle\langle 0|$ is the outer product of the bra $|0\rangle$ and ket $\langle 0|$, which, differs from the scalar product, results in a square matrix

$$|0\rangle\langle 0| = \begin{bmatrix} 1 \\ 0 \end{bmatrix} \cdot [1 \quad 0] = \begin{bmatrix} 1 & 0 \\ 0 & 0 \end{bmatrix}. \tag{2.13}$$

It should be noted that in general one can have four outer products using the computational basis, i.e., $|i\rangle\langle j|$ where $i, j = 0, 1$. Moreover, one can easily see that

$$|0\rangle\langle 0| + |1\rangle\langle 1| = \begin{bmatrix} 1 & 0 \\ 0 & 1 \end{bmatrix} = I, \tag{2.14}$$

where I is the identity matrix (or identity operator in Hilbert space).

The outer product $|i\rangle\langle i|$ is also called the projector onto $|i\rangle$ when it is operated to a ket on the left. It turns out $|i\rangle\langle i|$ can also operate to a bra on the right to project a quantum state onto the basis bra $\langle i|$. Therefore, any qubit $|\psi\rangle$ can be written in any basis by inserting an appropriate identity matrix. For example, in the Hadamard basis, $I = |+\rangle\langle +| + |-\rangle\langle -|$ and hence

$$|\psi\rangle = (|+\rangle\langle +| + |-\rangle\langle -|)|\psi\rangle = \langle +|\psi\rangle|+\rangle + \langle -|\psi\rangle|-\rangle. \tag{2.15}$$

Here

$$|\pm\rangle\langle\pm| = \left(\frac{|0\rangle \pm |1\rangle}{\sqrt{2}}\right)\left(\frac{\langle 0| \pm \langle 1|}{\sqrt{2}}\right) = \frac{|0\rangle\langle 0| + |1\rangle\langle 1| \pm |0\rangle\langle 1| \pm |1\rangle\langle 0|}{2}. \tag{2.16}$$

More generally, for a d-dimensional qudit (a quantum state that can assume d mutually exclusive events), the identity operator can be resolved as $I = \sum_{k=0}^{d-1} |k\rangle\langle k|$. A qubit corresponds to $d = 2$.

Finally, a convenient way to represent a qubit is the so-called Bloch sphere representation in the computational basis:

$$|\psi\rangle = \alpha|0\rangle + \beta|1\rangle = e^{i\chi}\left(\cos\frac{\theta}{2}|1\rangle + e^{i\varphi}\sin\frac{\theta}{2}|1\rangle\right), \tag{2.17}$$

with

$$\theta = 2\arctan\left|\frac{\beta}{\alpha}\right|, \quad \varphi = \arg\frac{\beta}{|\alpha|}, \quad \chi = \arg\alpha, \tag{2.18}$$

where $0 \le \theta \le \pi$ and $0 \le \varphi, \chi \le 2\pi$. The angle χ is a global phase which usually does not play any role. Figure 2.1 gives a pictorial illustration of a qubit on the Bloch sphere.

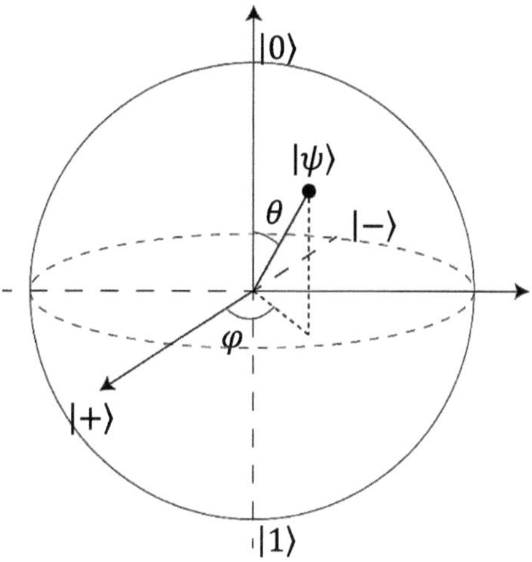

Fig. 2.1 Bloch sphere representation of a qubit $|\psi\rangle$

In this representation, the z-axis corresponds to the computational basis whereas the x-axis corresponds to the Hadamard basis. In fact, the qubit states along the x, y and z axes are the eigenkets of the X, Y and Z gates described in Sect. 2.2.3.

2.1.2 Multiple Qubits

The situation of multiple qubits is represented with the use of the tensor product. For example, a two-qubit system composed of the qubits $|\psi\rangle$ and $|\psi'\rangle$ is written as

$$|\psi\rangle|\psi'\rangle = |\psi\rangle \otimes |\psi'\rangle, \quad \text{or simply} \quad |\psi\psi'\rangle. \tag{2.19}$$

Here the qubits are distinguishable and the order of how the qubits are arranged matters: $|\psi\psi'\rangle = |\psi\rangle|\psi'\rangle \neq |\psi'\rangle|\psi\rangle = |\psi'\psi\rangle$. The left most qubit is conventionally referred to as the first qubit, followed by the second qubit on the right, and so on. When we want to avoid the ordering issue, we shall denote the qubit with a label explicitly, e.g., $|\psi\rangle^A|\psi'\rangle^B = |\psi'\rangle^B|\psi\rangle^A$. The distinguishability of the qubits can be attributed to the different degrees of freedom of the qubits, such as the spatial, temporal, spectral, or polarization modes. For quantum communication, a well-known example for the labels A and B are the separate spatial locations occupied by the two users, Alice and Bob. In the next section when multiple indistinguishable photons in the same mode are considered, we will introduce the notation of the Fock state.

The linear superposition of two-qubit states is also a two-qubit state. Therefore, in the computational basis, a two-qubit state can generally be written as

$$|\xi\rangle^{AB} = \alpha|00\rangle^{AB} + \beta|01\rangle^{AB} + \gamma|01\rangle^{AB} + \delta|11\rangle^{AB}, \tag{2.20}$$

where α, β, γ, and δ are complex numbers satisfying the normalization condition $|\alpha|^2 + |\beta|^2 + |\gamma|^2 + |\delta|^2 = 1$. It should be noted that Eq. (2.20) may not be written as a separable product in the form of Eq. (2.19). In such a circumstance, the state is said to be an entangled state. If a bipartite state can be written in a separable form, i.e., $|\xi\rangle^{AB} = |\psi\rangle^{A}|\psi'\rangle^{B}$ for some single qubit states $|\psi\rangle^{A}$ and $|\psi'\rangle^{B}$, the state is said to be separable or unentangled.

In quantum cryptography, one of the most important entangled two-qubit systems are the four Bell states [1]:

$$|\Phi^{+}\rangle^{AB} = |\beta_{00}\rangle^{AB} = \frac{|00\rangle^{AB} + |11\rangle^{AB}}{\sqrt{2}}, \tag{2.21a}$$

$$|\Phi^{-}\rangle^{AB} = |\beta_{10}\rangle^{AB} = \frac{|00\rangle^{AB} - |11\rangle^{AB}}{\sqrt{2}}, \tag{2.21b}$$

$$|\Psi^{+}\rangle^{AB} = |\beta_{01}\rangle^{AB} = \frac{|01\rangle^{AB} + |10\rangle^{AB}}{\sqrt{2}}, \tag{2.21c}$$

$$|\Psi^{+}\rangle^{AB} = |\beta_{11}\rangle^{AB} = \frac{|01\rangle^{AB} - |10\rangle^{AB}}{\sqrt{2}}. \tag{2.21d}$$

The Bell's states are the fundamental entangled resources for entanglement based quantum communication protocol such as quantum teleportation and the E91 QKD protocol.

On the other hand, given a two-qubit state in the form of Eq. (2.20) for some values α, β, γ, and δ, it may not be readily apparent to see whether the state is entangled or separable. One way to determine this is by the method of Schmidt decomposition [1]. The procedure is by writing the coefficients in a matrix and applying the singular-value decomposition (SVD):

$$M = \begin{bmatrix} \alpha & \beta \\ \gamma & \delta \end{bmatrix} = U\Sigma V^{\dagger},$$

where U and V are unitary matrices and $\Sigma = \begin{bmatrix} \lambda_1 & 0 \\ 0 & \lambda_2 \end{bmatrix}$ is a diagonal matrix. Then the bipartite state can be written as

$$|\xi\rangle^{AB} = \lambda_1|u_1\rangle^{A}|v_1\rangle^{B} + \lambda_2|u_2\rangle^{A}|v_2\rangle^{B}, \tag{2.22}$$

in which $|u_j\rangle, j = 1, 2$, are formed from the columns of U whereas $|v_j\rangle$ are from the columns of V. One can see that the state is separable if and only if only one of the eigenvalues of Σ is nonzero and has the value 1 (due to the normalization condition). On the other hand, $|\xi\rangle^{AB}$ is maximally entangled if $\lambda_1 = \lambda_2 = 1/\sqrt{2}$.

As an example, consider the bipartite state:

$$|\xi\rangle^{AB} = \frac{|00\rangle^{AB} + |01\rangle^{AB} + |10\rangle^{AB}}{\sqrt{3}}. \tag{2.23}$$

One can construct the matrix

$$M = \frac{1}{\sqrt{3}} \begin{bmatrix} 1 & 1 \\ 1 & 0 \end{bmatrix}. \tag{2.24}$$

To find the singular value decomposition of $M = U\Sigma V^\dagger$, one considers $MM^\dagger$ and its diagonalization

$$MM^\dagger = \frac{1}{3} \begin{bmatrix} 2 & 1 \\ 1 & 1 \end{bmatrix} = UDU^\dagger. \tag{2.25}$$

Note that D is positive definite and $D = \Sigma^2$. Then it is straightforward to obtain

$$D = \begin{bmatrix} \frac{1}{2} + \frac{\sqrt{5}}{6} & 0 \\ 0 & \frac{1}{2} - \frac{\sqrt{5}}{6} \end{bmatrix}, \quad \text{and} \quad U = \begin{bmatrix} \frac{3+\sqrt{5}}{2\sqrt{5+2\sqrt{5}}} & \frac{3-\sqrt{5}}{2\sqrt{5-2\sqrt{5}}} \\ \frac{1+\sqrt{5}}{2\sqrt{5+2\sqrt{5}}} & \frac{1-\sqrt{5}}{2\sqrt{5-2\sqrt{5}}} \end{bmatrix}, \tag{2.26}$$

hence

$$\Sigma = \sqrt{D} = \begin{bmatrix} \sqrt{\frac{1}{2} + \frac{\sqrt{5}}{6}} & 0 \\ 0 & \sqrt{\frac{1}{2} - \frac{\sqrt{5}}{6}} \end{bmatrix}. \tag{2.27}$$

Moreover,

$$V^\dagger = \Sigma^{-1} U^\dagger M = \begin{bmatrix} \frac{1+\sqrt{5}}{\sqrt{10+2\sqrt{5}}} & \sqrt{\frac{2}{5+\sqrt{5}}} \\ \frac{1-\sqrt{5}}{\sqrt{10-2\sqrt{5}}} & \sqrt{\frac{2}{5-\sqrt{5}}} \end{bmatrix}. \tag{2.28}$$

Using Σ, U, and V, the Schmidt decomposition of $|\xi\rangle^{AB}$ in Eq. (2.23) is given by

$$\lambda_1 = \sqrt{\frac{1}{2} + \frac{\sqrt{5}}{6}}, \quad \lambda_2 = \sqrt{\frac{1}{2} - \frac{\sqrt{5}}{6}},$$

and

$$|u_1\rangle = \frac{(3+\sqrt{5})|0\rangle + (1+\sqrt{5})|1\rangle}{2\sqrt{5+2\sqrt{5}}}, \quad |u_2\rangle = \frac{(3-\sqrt{5})|0\rangle + (1-\sqrt{5})|1\rangle}{2\sqrt{5-2\sqrt{5}}},$$
$$|v_1\rangle = \frac{(1+\sqrt{5})|0\rangle + 2|1\rangle}{\sqrt{10+2\sqrt{5}}}, \quad |v_2\rangle = \frac{(1-\sqrt{5})|0\rangle + 2|1\rangle}{\sqrt{10-2\sqrt{5}}}.$$

It should be noted that $\lambda_1 > \lambda_2$ in the example. The state $|\xi\rangle^{AB}$ in Eq. (2.23) is thus not maximally entangled.

As a final remark, a multiple qubit state can be analogously constructed using the tensor product. However, there is no simple way like the Schmidt decomposition to determine the separability of a quantum state involving more than 2 qubits.

2.1.3 *Qubit Operations*

In a closed quantum system, the evolution of the state of the system is reversible [1]. An open quantum system can be treated by considering an environment upon which one does not have control. The system + environment can hence still be considered as a closed system.

The reversibility of the evolution of a quantum system implies that an operation U on the system is unitary, i.e., $U^\dagger U = I$. In matrix notation, U is a unitary matrix. The evolved state by U can be written as

$$|\psi'\rangle = U|\psi\rangle. \tag{2.29}$$

It should be noted that

$$\langle\psi'|\psi'\rangle = (U|\psi\rangle)^\dagger(U|\psi\rangle) = \langle\psi|U^\dagger U|\psi\rangle = \langle\psi|\psi\rangle. \tag{2.30}$$

Hence the norm of the quantum state is unchanged as expected.

There are three commonly used elementary single qubit operations:

NOT gate

The NOT gate is defined by

$$X|i\rangle = |i \oplus 1\rangle, \tag{2.31}$$

where addition is modulo 2 for qubit. One can easily check that

$$X|0\rangle = |1\rangle, \quad X|1\rangle = |0\rangle, \quad X|\pm\rangle = \pm|\pm\rangle. \tag{2.32}$$

Hence the Hadamard basis is the eigenbasis of the NOT gate. In matrix notation,

$$X = \begin{bmatrix} 0 & 1 \\ 1 & 0 \end{bmatrix}, \quad X|\psi\rangle = X\begin{bmatrix} \alpha \\ \beta \end{bmatrix} = \begin{bmatrix} \beta \\ \alpha \end{bmatrix}. \tag{2.33}$$

Z gate (Phase-flip operator)

The Z gate is defined similarly by

$$Z|0\rangle = |0\rangle, \quad Z|1\rangle = -|1\rangle. \tag{2.34}$$

Hence the computational basis is the eigenbasis of the Z gate with the eigenvalues ± 1. One can easily check that

$$Z|+\rangle = |-\rangle, \quad Z|-\rangle = |+\rangle. \tag{2.35}$$

In matrix notation,

$$Z = \begin{bmatrix} 1 & 0 \\ 0 & -1 \end{bmatrix}, \quad Z|\psi\rangle = Z\begin{bmatrix} \alpha \\ \beta \end{bmatrix} = \begin{bmatrix} \alpha \\ -\beta \end{bmatrix}. \tag{2.36}$$

That is, the Z gate flips the phase of the $|1\rangle$ component with respect to the $|0\rangle$ component.

The X and Z gates can be combined to form the Y gate: $Y = iXZ$, where

$$Y = \begin{bmatrix} 0 & -i \\ i & 0 \end{bmatrix}. \tag{2.37}$$

The three matrices X, Y, and Z are also identified to be the three Pauli matrices.

Hadamard gate

The Hadamard gate is defined by

$$H|0\rangle = |+\rangle, \quad H|1\rangle = |-\rangle. \tag{2.38}$$

Hence the Hadamard gate connects the computational basis and the Hadmard basis. One can easily check that

$$H = |+\rangle\langle 0| + |-\rangle\langle 1| = \frac{X+Z}{\sqrt{2}}. \tag{2.39}$$

In matrix notation,

$$H = \frac{1}{\sqrt{2}}\begin{bmatrix} 1 & 1 \\ 1 & -1 \end{bmatrix}. \tag{2.40}$$

For multiple qubits, we can also define multiple qubit operations by means of the tensor product, e.g., $U \otimes U' \otimes U'' = UU'U''$. Again, the order of the operations matters as we assume that the qubits are distinguishable. To avoid the ordering issue, we can likewise use labels to denote which operator apply on which qubit. For example,

$$\begin{aligned} X^A Z^B Z^C H^D |\psi\phi\chi\mu\rangle^{ABCD} &= X^A Z^B Z^C H^D |\psi\rangle^A |\phi\rangle^B |\chi\rangle^C |\mu\rangle^D \\ &= Z^C |\chi\rangle^C H^D |\mu\rangle^D X^A |\psi\rangle^A Z^B |\phi\rangle^B, \end{aligned} \tag{2.41}$$

where we have taken the freedom to move around the operators and states as long as the operators are acting on the left of the corresponding kets.

In addition to single qubit operations, one can also define two-qubit operations. In particular, the most useful two-qubit operation is the controlled-NOT (CNOT) gate defined by:

$$U_{\text{CNOT}}^{AB} = |0\rangle\langle 0|^A \otimes I^B + |1\rangle\langle 1|^A \otimes X^B. \tag{2.42}$$

In matrix notation, it reads

$$U_{\text{CNOT}}^{AB} = \begin{bmatrix} 1 & 0 \\ 0 & 0 \end{bmatrix} \otimes \begin{bmatrix} 1 & 0 \\ 0 & 1 \end{bmatrix} + \begin{bmatrix} 0 & 0 \\ 0 & 1 \end{bmatrix} \otimes \begin{bmatrix} 0 & 1 \\ 1 & 0 \end{bmatrix} = \begin{bmatrix} 1 & 0 & 0 & 0 \\ 0 & 1 & 0 & 0 \\ 0 & 0 & 0 & 1 \\ 0 & 0 & 1 & 0 \end{bmatrix}. \tag{2.43}$$

In fact, the Bell states can be created using U_{CNOT}^{AB} and the Hadamard gate:

$$|\Phi^+\rangle^{AB} = U_{\text{CNOT}}^{AB} H^A |0\rangle^A |0\rangle^B. \tag{2.44}$$

2.1.4 Mixed States and Density Operators

The ket considered in the previous sections are called *pure states* in quantum mechanics, of which the probability amplitudes in the state corresponds to pure quantum randomness of a closed quantum system. On the other hand, there are situations in which the indeterminacy of the system is due to the lack of knowledge about the system. For example, it could happen that a quantum system can be in either one of two states $|\psi_0\rangle$ or $|\psi_1\rangle$ randomly according to some known or unknown probability distribution $p_X(x)$, $x \in \mathcal{X} = \{0, 1\}$, $\sum_x p_X(x) = 1$. The system is called to be in a mixed state denoted by the ensemble $\mathcal{E} = \{p_X(x), |\psi_x\rangle\}_{x\in\mathcal{X}}$.

In this case, one expects that the overall quantum state of the system should be an incoherent sum of the states $|\psi_0\rangle$ and $|\psi_1\rangle$, weighted by p_x. In fact, the quantum state is described by the density operator (also called density matrix)

$$\rho = \sum_{x \in \mathcal{X}} p_X(x) |\psi_x\rangle\langle\psi_x|. \tag{2.45}$$

In the density operator formalism, a pure state can be written as $\rho = |\psi\rangle\langle\psi|$. That is, the density operator ρ in Eq. (2.45) represents a pure state if there is only a single term in the sum, otherwise it is a mixed state. Note that $\{|\psi_x\rangle\}$ are normalized but need not be orthogonal to each other.

It can be seen that the density operator is a sum of the outer products of the states $|\psi_x\rangle$ in the ensemble $\mathcal{E}$ weighted by $p_X(x)$. Hence ρ is Hermitian, i.e., $\rho = \rho^\dagger$, and can be written in a diagonalized form $\rho = \sum_k \lambda_k |\phi_k\rangle\langle\phi_k|$, with $\{|\phi_k\rangle\}$ being an orthonormal set of kets. The evolution of ρ by the unitary operator U is given by [1]

$$\rho' = U \rho U^\dagger. \tag{2.46}$$

In matrix theory, the trace of a square matrix M is defined as the sum of the diagonal matrix elements of M, i.e., $\mathrm{Tr}\, M = \sum_j \langle j|M|j\rangle = \sum_j M_{jj}$, where $\{|j\rangle\}$ is some set of basis kets and $\{M_{ij}\}$ are the matrix elements of M represented in the basis $\{|j\rangle\}$. Since quantum states are normalized, it can be shown that the trace of a density operator is one:

$$\begin{aligned}
\mathrm{Tr}\, \rho &= \sum_j \langle j|\rho|j\rangle = \sum_{x \in \mathcal{X}} p_X(x) \sum_j \langle j|\psi_x\rangle\langle\psi_x|j\rangle \\
&= \sum_{x \in \mathcal{X}} p_X(x) \sum_j \langle\psi_x|j\rangle\langle j|\psi_x\rangle = \sum_{x \in \mathcal{X}} p_X(x) \langle\psi_x|\psi_x\rangle \\
&= \sum_{x \in \mathcal{X}} p_X(x) = 1,
\end{aligned} \tag{2.47}$$

in which the identity $I = \sum_j |j\rangle\langle j|$ is used to get rid of the arbitrary basis $\{|j\rangle\}$. Using the diagonalized form of ρ, it can also be found that the sum of the eigenvalues $\sum_k \lambda_k = 1$. That is, the set of the eigenvalues and eigenkets $\{\lambda_k, |\phi_k\rangle\}$ can be regarded as an ensemble of pure states for the density operator ρ. In general, a given density operator can be represented by different ensembles of pure states. The trace operation can also be used to determine the purity, defined as $\mathrm{Tr}\, \rho^2$, of the density operator. The purity can be proved to have a value between zero and one by using the diagonalized form of ρ. If the state is a pure state, the purity is equal to one, and vice versa.

For multipartite quantum states, the trace operation provides the mechanism to determine the quantum states of the subsystems. Specifically, for the bipartite quantum state $\rho^{AB} = \sum_{x \in \mathcal{X}} p_X(x)|\psi_x\rangle^{AB}\langle\psi_x|$, the quantum state of the subsystem A is given by

$$\rho^A = \mathrm{Tr}_B \rho^{AB} = \sum_{x \in \mathcal{X}} p_X(x) \mathrm{Tr}_B\left(|\psi_x\rangle^{AB}\langle\psi_x|\right) = \sum_{x \in \mathcal{X}} p_X(x)\rho_x^A, \tag{2.48}$$

where the trace is applied to the subsystem B only (hence it is also called partial trace in this case). In particular, for a pure state $|\psi\rangle^{AB}$, the density operator for A is $\rho^A = \mathrm{Tr}_B\left(|\psi\rangle^{AB}\langle\psi|\right)$. It should be noted in Eq. (2.48) that although ρ^A is written as an incoherent sum of the set of density operators $\{\rho_x^A\}$, ρ^A can also always be recast as an incoherent sum of a set of pure states, for example, using the eigenkets of ρ^A.

One of the main applications of the partial trace is on the situation when the subsystem A corresponds to a quantum system that interacts with the environment E. Given the system density operator $\rho^A = \sum_{x \in \mathcal{X}} p_X(x)|\psi_x\rangle^A\langle\psi_x|$, one can always construct a composite pure state for the system + environment $|\psi\rangle^{AE}$ such that $\rho^A = \mathrm{Tr}_E\left(|\psi\rangle^{AE}\langle\psi|\right)$. Explicitly, the composite state is

$$|\psi\rangle^{AE} = \sum_{x \in \mathcal{X}} \sqrt{p_X(x)}|\psi_x\rangle^A|\phi_x\rangle^E, \tag{2.49}$$

where $\{|\phi_x\rangle^E\}$ is some orthonormal basis for the environment E. Equation (2.49) is called the purification of ρ^A on the reference system (or environment) E.

The concept of purification can be applied to understand general quantum operations. As mentioned at the beginning of this subsection, an open quantum system can be regarded as a closed system comprising of the quantum system and the environment. As a consequence, the quantum operation $\mathcal{E}$ on the system ρ^A can be seen as equivalent to a unitary operation U^{AE} on the system + environment composite:

$$\rho^A \to \mathcal{E}(\rho^A) = \mathrm{Tr}_E\left[U^{AE}(\rho^A \otimes \rho^E)U^{AE\dagger}\right], \tag{2.50}$$

where ρ^E is the state of the environment that is *assumed to be separable from the system*. We shall not go into the implication of Eq. (2.50). For details, the readers are referred to [1].

2.1.5 No-Cloning Theorem

One of the most important foundation of quantum cryptography is the no-cloning theorem [3, 4]. In classical communication, information encoded in bits can in principle be copied perfectly. However, this may not be valid when the information is encoded in qubits. In a rigorous term, the no-cloning theorem says that it is impossible to build a universal copier of quantum states, in which the universal copier is taken to be a unitary operation. Mathematically, this theorem can be proved using contradiction.

Suppose there exists an unitary U such that

$$U|\psi\rangle|0\rangle = |\psi\rangle|\psi\rangle = |\psi\psi\rangle, \tag{2.50a}$$

$$U|\phi\rangle|0\rangle = |\phi\rangle|\phi\rangle = |\phi\phi\rangle, \tag{2.50b}$$

where $|\psi\rangle$ and $|\phi\rangle$ are some arbitrary quantum states. Here $|0\rangle$ is the initial state of the copier. Consider the scalar product between the right-hand sides of Eq. (2.50):

$$\langle\phi\phi|\psi\psi\rangle = \langle\phi|\psi\rangle\langle\phi|\psi\rangle = \langle\phi|\psi\rangle^2.$$

On the other hand, the left-hand side gives,

$$\langle\phi|\langle 0|U^\dagger U|\psi\rangle|0\rangle = (\langle\phi|\langle 0|)(|\psi\rangle|0\rangle) = \langle\phi|\psi\rangle\langle 0|0\rangle = \langle\phi|\psi\rangle,$$

in which we have used the relations $U^\dagger U = I$ and $\langle 0|0\rangle = 1$. As a consequence, we get

$$\langle\phi|\psi\rangle(\langle\phi|\psi\rangle - 1) = 0 \quad \Rightarrow \quad \langle\phi|\psi\rangle = 0 \,\text{or}\, \langle\phi|\psi\rangle = 1. \tag{2.51}$$

Therefore, if U can clone ψ, it can only clone ϕ that is orthogonal to ψ if ϕ is different from ψ. Hence U is not universal.

The implication of the no-cloning theorem is that one cannot copy a quantum state perfectly if he does not know what the quantum state is. The well-known BB84 QKD protocol exactly exploits this fact by sending out the bit information encoded in qubit represented in different bases randomly [5]. We shall come to that in the next chapter.

The proof above applies to the case of pure states. It can be generalized to mixed states as well, which is also called the no-broadcasting theorem [6]. It should be noted that the no-cloning theorem asserts that an unknown quantum state cannot be duplicated exactly. However, an arbitrary state can still be *approximately* cloned [7–11].

2.1.6 *Quantum Measurement*

We understand the world by observing what it is and how it evolves. Therefore, to complete the description of quantum mechanics, we need to specify how the quantum states are measured. Quantum mechanics postulate that measurements are described by a collection of measurement operators $\{M_k\}$ that operate on the quantum state. The index k refers to the outcome of the measurement operator M_k. The set of measurement operators $\{M_k\}$ for a specific measurement task should be complete, in the sense that the measurement task always comes up with something given the quantum state $|\psi\rangle$. For example, for the measurement task of determining the whether a qubit is in the 0 or 1 state, the measurement operators can be $\{M_0, M_1\}$, where

$$M_0 = |0\rangle\langle 0|, M_1 = |1\rangle\langle 1|. \tag{2.52}$$

Note that M_0 and M_1 here are just the projectors for the states $|0\rangle$ and $|1\rangle$.

Given the measurement operators, a measurement M_k acts on the quantum state $|\psi\rangle$ and gives rise the post-measurement state:

$$|\psi'_f\rangle = M_k|\psi\rangle. \tag{2.53}$$

It should be noted that the post-measurement state in general no longer is normalized. In fact, its norm gives the probability p_k for the event k, where

$$p_k = \langle\psi'_f|\psi'_f\rangle = \langle\psi|M_k^\dagger M_k|\psi\rangle. \tag{2.54}$$

After the measurement, we should be certain that the quantum state is $|\psi'_f\rangle$, because the measurement outcome is k. Therefore, in order to properly described the post-measurement state, we can give it a proper normalization using p_k, viz.,

$$|\psi_f\rangle = \frac{M_k|\psi\rangle}{\sqrt{p_k}} = \frac{M_k|\psi\rangle}{\sqrt{\langle\psi|M_k^\dagger M_k|\psi\rangle}}. \tag{2.55}$$

Finally, since the measurement operators form a complete set for the measurement task, we should have

$$1 = \sum_k p_k = \sum_k \langle\psi|M_k^\dagger M_k|\psi\rangle = \langle\psi|\sum_k M_k^\dagger M_k|\psi\rangle. \tag{2.56}$$

This is true for any $|\psi\rangle$, hence the completeness of $\{M_k\}$ is equivalently represented by

$$\sum_k M_k^\dagger M_k = I. \tag{2.57}$$

For example, it can be easily seen that the measurement operators $\{M_0, M_1\}$ in Eq. (2.52) satisfies the completeness relation (2.57), and $M_k^\dagger M_k = M_k M_k = M_k$ for $k = 0, 1$. For a given density matrix ρ, Eqs. (2.54) and (2.55) are generalized as

$$p_k = \mathrm{Tr}\left(\rho M_k^\dagger M_k\right) = \mathrm{Tr}\left(M_k \rho M_k^\dagger\right), \tag{2.57a}$$

$$\rho_f = \frac{M_k \rho M_k^\dagger}{\mathrm{Tr}\left(M_k \rho M_k^\dagger\right)}. \tag{2.57b}$$

Finally, in quantum information, the notion of positive-operator-valued measure (POVM) is often used. In simple terms, a POVM is a set of operators $\{E_k\}$ that are positive ($E_k > 0$, i.e., E_k has positive eigenvalues) and complete $\left(\sum_k E_k = I\right)$. The connection between E_k and M_k is

$$E_k = M_k^\dagger M_k, \tag{2.58a}$$

or more generally

$$E_k = \sum_l M_{k,l}^\dagger M_{k,l}, \tag{2.58b}$$

for some subset $\{M_{k,l}\}_l$ of the full set $\{M_{k,l}\}_{k,l}$. The positivity and completeness of $\{E_k\}$ can be deduced from Eq. (2.58b). The POVM is introduced because often time one is interested in the measurement probability p_k only and do not care about the post-measurement state. In this way, the probability is simply given by the expectation value

$$P(k|\rho) = p_k = \mathrm{Tr}(\rho E_k), \tag{2.59}$$

in which we have explicitly use the conditional probability $P(k|\rho)$ to denote the probability of the outcome k given the density matrix ρ. Also it is usually easier to construct a POVM than the measurement operators $\{M_k\}$. For example, we can construct a measurement task to distinguish between $|0\rangle$ and $|+\rangle$. Note that $|0\rangle$ and $|+\rangle$ are nonorthogonal. It turns out that the following POVM can unambiguously distinguish these two qubits:

$$E_1 = a|1\rangle\langle 1|, \tag{2.60a}$$

$$E_2 = a|-\rangle\langle -|, \tag{2.60b}$$

$$E_3 = I - E_1 - E_2, \tag{2.60c}$$

where $a = \frac{\sqrt{2}}{1+\sqrt{2}}$. Note that E_1 and E_2 are constructed by using the complements of $|0\rangle$ and $|+\rangle$ respectively, whereas E_3 is needed in order to satisfy the completeness condition. The coefficient a is chosen so that E_k's are positive and E_3 can be written as a projector, viz.

$$E_3 = \frac{1}{\sqrt{2}}\left[|0\rangle - \left(\sqrt{2}-1\right)|1\rangle\right]\left[\langle 0| - \left(\sqrt{2}-1\right)\langle 1|\right].$$

Therefore $\langle 0|E_1|0\rangle = \langle +|E_2|+\rangle = 0$ and $\langle +|E_1|+\rangle$, $\langle 0|E_2|0\rangle$, $\langle 0|E_3|0\rangle$, $\langle +|E_3|+\rangle$ are nonzero. As a result, $|0\rangle$ and $|+\rangle$ can be distinguished without misidentification. For example, given $|\psi\rangle = |0\rangle$ or $|+\rangle$, if the measurement outcome is E_1, the state $|\psi\rangle$ must have been $|+\rangle$ because this nonzero measurement outcome cannot come from $|0\rangle$ (i.e., $\langle 0|E_1|0\rangle = 0$). Likewise, if the outcome is E_2, the state must have been $|0\rangle$. However, if the outcome is E_3, the measurement is inconclusive. This example illustrates the unambiguous state discrimination of two nonorthogonal states.

2.2 Quantum Theory of Photons

In the previous section, we briefly mentioned a qubit realized by the polarization states of light. In fact, the first quantum key distribution protocol, the BB84 protocol, was proposed using the polarization of photons [5]. In this section, we shall review the physical properties of light and the concept of photons through the quantization of the electromagnetic field. In particular, the aspects of which that can be conceptually and nicely associated to that of a qubit will be described in detail. This chapter also sets the stage of the description of multi-photon states for the discussion of multi-photon quantum communications in the rest of the book.

2.2.1 *Quantization of Electromagnetic Field*

According to classical electromagnetic (EM) theory, light (in the usual sense of visible light with wavelength from 400 to 700 nm) is a form of EM wave that satisfies the wave equation. For a plane EM wave of frequency ω_k propagating in the direction of the wave vector $\vec{k}$ in free space, the electric and magnetic fields can be written as [12]

$$\vec{E}(\vec{r},t) = \vec{\epsilon}E_0 e^{i\left(\vec{k}\cdot\vec{r}-\omega_k t\right)} + c.c., \tag{2.61a}$$

$$\vec{B}(\vec{r},t) = \frac{\vec{k}\times\vec{\epsilon}}{kc}E_0 e^{i\left(\vec{k}\cdot\vec{r}-\omega_k t\right)} + c.c., \tag{2.61b}$$

where $k = \left|\vec{k}\right| = \omega_k/c$ with the speed of light c and $\vec{\epsilon}$ is the unit polarization vector of the EM field. The notation $c.c.$ stands for the complex conjugate. It should be noted that the divergence equations $\nabla\cdot\vec{E}(\vec{r},t) = 0$ and $\nabla\cdot\vec{B}(\vec{r},t) = 0$ demand that both $\vec{E}$ and $\vec{B}$ are perpendicular to the propagation direction $\vec{k}$. Therefore, the polarization vector $\vec{\epsilon}$ lies on the two-dimensional transverse plane that is orthogonal to $\vec{k}$. We denote the two orthogonal polarizations by $\vec{\epsilon}_{\vec{k}s}$ with $s = 1, 2$ for the given $\vec{k}$. For an EM wave propagating in the z direction, the polarization vector can be written as

$$\vec{\epsilon}_{\vec{k}s} = \epsilon_{xs}\hat{x} + \epsilon_{ys}\hat{y}, \tag{2.62}$$

with $|\epsilon_{xs}|^2 + |\epsilon_{ys}|^2 = 1$ for both $s = 1, 2$. In general, ϵ_{xs} and ϵ_{ys} are complex numbers. Some special cases are shown in Table 2.1.

It is apparent that the two-dimensional nature of the polarization vectors enables them to represent the two-dimensional qubit discussed in the previous section. One can also immediately associate the computational basis $\{|0\rangle, |1\rangle\}$ to the horizontal and vertical polarizations, and the Hadamard basis $\{|+\rangle, |-\rangle\}$ to the diagonal and anti-diagonal polarizations.

It should be remarked that up till now we are still treating classical EM fields. In order to describe the corresponding quantum version and to introduce the concept of photons, the procedure is to carry out the quantization of the field [13, 14]. We shall only focus on the quantization of the EM field in free space.

To begin, we first represent the electric and magnetic fields using the potentials. We shall use the Coulomb gauge or transverse gauge so that the scalar potential is set to zero and the vector potential $\vec{A}(\vec{r}, t)$ satisfies the relation [12]

$$\nabla \cdot \vec{A}(\vec{r}, t) = 0. \tag{2.63}$$

The electric and magnetic fields are then written in terms of the vector potential as

$$\vec{E}(\vec{r}, t) = -\frac{\partial}{\partial t}\vec{A}(\vec{r}, t), \quad \vec{B}(\vec{r}, t) = \nabla \times \vec{A}(\vec{r}, t). \tag{2.64}$$

For a plane wave in free space, the vector potential is then written as

$$\vec{A}(\vec{r}, t) = \vec{\epsilon} A_0 e^{i(\vec{k}\cdot\vec{r} - \omega_k t)} + c.c., \tag{2.65}$$

where $A_0 = E_0/i\omega_k$ by means of Eq. (2.61a, b).

Table 2.1 Examples of polarization states and the corresponding polarization vectors

Polarization states	Polarization vectors
Horizontal (H) and vertical (V) polarizations	$\vec{\epsilon}_{\vec{k}1} = \vec{x}, \quad \vec{\epsilon}_{\vec{k}2} = \vec{y}$
Diagonal (D) and anti-diagonal (A) polarizations	$\vec{\epsilon}_{\vec{k}1} = \frac{1}{\sqrt{2}}(\vec{x} + \vec{y}), \quad \vec{\epsilon}_{\vec{k}2} = \frac{1}{\sqrt{2}}(\vec{x} - \vec{y})$
Left (L) and right (R) circular polarizations	$\vec{\epsilon}_{\vec{k}1} = \frac{1}{\sqrt{2}}(\vec{x} + i\vec{y}), \quad \vec{\epsilon}_{\vec{k}2} = \frac{1}{\sqrt{2}}(\vec{x} - i\vec{y})$

Now we consider an electromagnetic field confined within the domain of a box with volume $V = L^3$ and expand the field in terms of a complete set of plane waves in the box. By the boundary conditions on the surface of the box, this set of plane waves takes on certain values of the wave vectors only:

$$\vec{k} = \frac{2\pi\vec{n}}{L}, \tag{2.66}$$

where $\vec{n} = (n_x, n_y, n_z)$ with $n_j = 0, \pm 1, \pm 2, \ldots$. The vector potential in the box is then given in terms of the plane wave expansion by

$$\vec{A}(\vec{r}, t) = \frac{1}{\sqrt{V}} \sum_{\vec{k}s} \vec{\epsilon}_{\vec{k}s} A_{\vec{k}s} e^{i\left(\vec{k}\cdot\vec{r} - \omega_k t\right)} + c.c., \tag{2.67}$$

where the summation is over all allowed wave vector values in the box and the two orthogonal polarizations.

In quantum theory, the fields $\vec{E}$, $\vec{B}$ and $\vec{A}$ become operators. We introduce the notation

$$a_{\vec{k}s} = \sqrt{\frac{2\omega_k \epsilon_0}{\hbar}} A_{\vec{k}s}, \tag{2.68}$$

where $\hbar$ is the reduced Planck constant and ϵ_0 is the permittivity of vacuum, such that the vector potential is expressed as

$$\vec{A}(\vec{r}, t) = \sum_{\vec{k}s} \sqrt{\frac{\hbar}{2\omega_k \epsilon_0 V}} \vec{\epsilon}_{\vec{k}s} a_{\vec{k}s} e^{i\left(\vec{k}\cdot\vec{r} - \omega_k t\right)} + h.c., \tag{2.69}$$

where $h.c.$ denotes the Hermitian conjugate. According to quantum theory, the quantization of the EM field is established by requiring $a_{\vec{k}s}$ and $a^{\dagger}_{\vec{k}s}$ to be operators that satisfy the following relation [13, 14]

$$\left[a_{\vec{k}s}, a^{\dagger}_{\vec{k}'s'}\right] = \delta_{\vec{k}\vec{k}'}\delta_{ss'}, \quad \left[a_{\vec{k}s}, a_{\vec{k}'s'}\right] = 0, \tag{2.70}$$

in which $[A, B] \equiv AB - BA$ is called the commutator of A and B. The operators $a_{\vec{k}s}$ and $a^{\dagger}_{\vec{k}s}$ are called the annihilation and creation operators respectively. The commutation relations (2.70) take their form through the interpretation that the quantized EM field in free space behaves like a collection of harmonic oscillators (see Eq. (2.73) below).

Using the annihilation and creation operators, the electric and magnetic operators can likewise be expressed explicitly:

$$\vec{E}(\vec{r},t) = -i\sum_{\vec{k}s}\sqrt{\frac{\hbar\omega_k}{2\epsilon_0 V}}\left[\vec{\epsilon}_{\vec{k}s} a_{\vec{k}s} e^{i\left(\vec{k}\cdot\vec{r}-\omega_k t\right)} - \vec{\epsilon}^{\,*}_{\vec{k}s} a^{\dagger}_{\vec{k}s} e^{-i\left(\vec{k}\cdot\vec{r}-\omega_k t\right)}\right], \tag{2.71a}$$

$$\vec{B}(\vec{r},t) = i\sum_{\vec{k}s}\sqrt{\frac{\hbar\omega_k}{2\epsilon_0 c^2 V}}\vec{\kappa}\times\left[\vec{\epsilon}_{\vec{k}s} a_{\vec{k}s} e^{i\left(\vec{k}\cdot\vec{r}-\omega_k t\right)} - \vec{\epsilon}^{\,*}_{\vec{k}s} a^{\dagger}_{\vec{k}s} e^{-i\left(\vec{k}\cdot\vec{r}-\omega_k t\right)}\right], \tag{2.71b}$$

where $\vec{\kappa} = \vec{k}/k$. By using the orthogonality relation of the plane waves

$$\frac{1}{V}\int_V e^{i\left(\vec{k}\cdot\vec{r}-\vec{k}'\cdot\vec{r}\right)} d^3r = \delta_{n_x n'_x}\delta_{n_y n'_y}\delta_{n_z n'_z} = \delta_{\vec{k}\vec{k}'}, \tag{2.72}$$

and noticing that

$$\int_V d\vec{r}\left|\vec{E}(\vec{r},t)\right| = \frac{\hbar}{2\epsilon_0}\sum_{\vec{k}ss'}\omega_k\left[\delta_{ss'} a_{\vec{k}s} a^{\dagger}_{\vec{k}s} - \vec{\epsilon}_{\vec{k}s}\cdot\vec{\epsilon}_{-\vec{k}s'} a_{\vec{k}s} a_{-\vec{k}s'} e^{-2i\omega_k t} + h.c.\right]$$

$$\begin{aligned}\int_V d\vec{r}\left|\vec{B}(\vec{r},t)\right| &= \frac{\hbar\mu_0}{2}\sum_{\vec{k}ss'}\omega_k\left[\left(\vec{\kappa}\times\vec{\epsilon}_{\vec{k}s}\right)\cdot\left(\vec{\kappa}\times\vec{\epsilon}^{\,*}_{\vec{k}s'}\right) a_{\vec{k}s} a^{\dagger}_{\vec{k}s'}\right.\\ &\quad\left. + \left(\vec{\kappa}\times\vec{\epsilon}_{\vec{k}s}\right)\cdot\left(\vec{\kappa}\times\vec{\epsilon}_{-\vec{k}s'}\right) a_{\vec{k}s} a_{-\vec{k}s'} e^{-2i\omega_k t} + h.c.\right]\\ &= \frac{\hbar\mu_0}{2}\sum_{\vec{k}ss'}\omega_k\left[\delta_{ss'} a_{\vec{k}s} a^{\dagger}_{\vec{k}s} + \vec{\epsilon}_{\vec{k}s}\cdot\vec{\epsilon}_{-\vec{k}s'} a_{\vec{k}s} a_{-\vec{k}s'} e^{-2i\omega_k t} + h.c.\right]\end{aligned}$$

the energy of the EM field inside the box can be found to be

$$\begin{aligned}\mathcal{H} &= \frac{1}{2}\int_V\left[\epsilon_0\left|\vec{E}(\vec{r},t)\right|^2 + \frac{1}{\mu_0}\left|\vec{B}(\vec{r},t)\right|^2\right] d^3r = \sum_{\vec{k}s}\frac{\hbar\omega_k}{2}\left(a_{\vec{k}s} a^{\dagger}_{\vec{k}s} + a^{\dagger}_{\vec{k}s} a_{\vec{k}s}\right)\\ &= \sum_{\vec{k}s}\hbar\omega_k\left(a^{\dagger}_{\vec{k}s} a_{\vec{k}s} + \frac{1}{2}\right),\end{aligned} \tag{2.73}$$

That is, the total energy is the sum of the energies of all the modes with wave vector $\vec{k}$ and polarization s.

The total energy of the EM field in Eq. (2.73) is also known as the Hamiltonian operator (or simply the Hamiltonian) of the quantized EM field. One can recognizes that $\mathcal{H}$ takes the same form as the Hamiltonian of a collection of quantized harmonic oscillators [2]. The operator $N_{\vec{k}s} \equiv a^{\dagger}_{\vec{k}s} a_{\vec{k}s}$ is called the number operator of the mode $\vec{k}s$. Together with the annihilation and creation operators, the quantum states of the EM field can be described explicitly as explained in the next section.

2.2.2 Photon States

The number operator is a positive semi-definite operator, i.e., given any quantum state $|\psi\rangle$, $\langle\psi|N_{\vec{k}s}|\psi\rangle = \langle\psi|a^{\dagger}_{\vec{k}s}a_{\vec{k}s}|\psi\rangle = \left\|a_{\vec{k}s}|\psi\rangle\right\|^2 \geq 0$. This is consistent with the understanding that the EM field energy is non-negative according to Eq. (2.73).

In the following, let us first consider the situation of a fixed mode $\vec{k}s$. The eigen-equation of $N_{\vec{k}s}$ reads as

$$N_{\vec{k}s}|\psi\rangle_{\vec{k}s} = \lambda|\psi\rangle_{\vec{k}s}, \tag{2.74}$$

where $\lambda \geq 0$ is the eigenvalue and $|\psi\rangle_{\vec{k}s}$ is the corresponding eigenket. Using the commutation relations (2.70), we find that

$$N_{\vec{k}s}a_{\vec{k}s}|\psi\rangle_{\vec{k}s} = a^{\dagger}_{\vec{k}s}a_{\vec{k}s}a_{\vec{k}s}|\psi\rangle_{\vec{k}s} = \left(a_{\vec{k}s}a^{\dagger}_{\vec{k}s}a_{\vec{k}s} - a_{\vec{k}s}\right)|\psi\rangle_{\vec{k}s} = (\lambda - 1)a_{\vec{k}s}|\psi\rangle_{\vec{k}s}. \tag{2.75}$$

Therefore, $a_{\vec{k}s}|\psi\rangle_{\vec{k}s}$ is also an eigenket of $N_{\vec{k}s}$ with the eigenvalue $(\lambda - 1)$. In general, it can be proved that, if $|\psi\rangle_{\vec{k}s}$ is an eigenket of $N_{\vec{k}s}$ with the eigenvalue $\lambda \geq 0$, the state

$$\left(a_{\vec{k}s}\right)^n|\psi\rangle_{\vec{k}s}, \tag{2.76}$$

is also an eigenket of $N_{\vec{k}s}$ with the eigenvalue $\lambda - n$, where n is an arbitrary nonnegative integer. However, since the number operator is proportional to the energy of the EM field, to make physical sense, we expect that λ should be finite. Since $\lambda - n \geq 0$ as it is an eigenvalue of $N_{\vec{k}s}$ and $|\psi\rangle_{\vec{k}s} \neq 0$, the only possibility is that λ is an integer so that $\lambda - n$ terminates at some finite integer n. In particular, the eigenket $|0\rangle_{\vec{k}s} = \left(a_{\vec{k}s}\right)^{\lambda}|\psi\rangle_{\vec{k}s}$ gives a zero eigenvalue for $N_{\vec{k}s}$:

$$N_{\vec{k}s}|0\rangle_{\vec{k}s} = 0. \tag{2.77}$$

The eigenket $|0\rangle_{\vec{k}s}$, taken to be normalized, is called the ground state of the EM field mode $\vec{k}s$. From Eq. (2.73),

$$\hbar\omega_k\left(a^{\dagger}_{\vec{k}s}a_{\vec{k}s} + \frac{1}{2}\right)|0\rangle_{\vec{k}s} = \frac{\hbar\omega_k}{2}|0\rangle_{\vec{k}s}. \tag{2.78}$$

Therefore, in the ground state, the EM field in the mode $\vec{k}s$ has the energy $\hbar\omega_k/2$, which is called the vacuum energy of the mode $\vec{k}s$.

On the other hand, consider the application of the creation operator to Eq. (2.77):

$$0 = a^{\dagger}_{\vec{k}s} N_{\vec{k}s} |0\rangle_{\vec{k}s} = a^{\dagger}_{\vec{k}s} a^{\dagger}_{\vec{k}s} a_{\vec{k}s} |0\rangle_{\vec{k}s} = \left(a^{\dagger}_{\vec{k}s} a_{\vec{k}s} a^{\dagger}_{\vec{k}s} - a^{\dagger}_{\vec{k}s} \right) |0\rangle_{\vec{k}s}. \tag{2.79}$$

Therefore,

$$N_{\vec{k}s} a^{\dagger}_{\vec{k}s} |0\rangle_{\vec{k}s} = a^{\dagger}_{\vec{k}s} |0\rangle_{\vec{k}s}, \tag{2.80}$$

that is, $a^{\dagger}_{\vec{k}s}|0\rangle_{\vec{k}s}$ is also an eigenket of $N_{\vec{k}s}$ with the eigenvalue 1. The corresponding normalized eigenstate

$$|1\rangle_{\vec{k}s} = a^{\dagger}_{\vec{k}s} |0\rangle_{\vec{k}s}, \tag{2.81}$$

is called the first excitation of the field mode $\vec{k}s$, or the single photon state of mode $\vec{k}s$. Applying the creation operator n times to $|0\rangle_{\vec{k}s}$, one obtains the n-th excitation, or the n-photon state, of mode $\vec{k}s$, which when properly normalized is given by

$$|n\rangle_{\vec{k}s} = \frac{1}{\sqrt{n!}} \left(a^{\dagger}_{\vec{k}s} \right)^n |0\rangle_{\vec{k}s}. \tag{2.82}$$

In fact, the n and $n-1$ photon states are related to each other by

$$a^{\dagger}_{\vec{k}s} |n-1\rangle_{\vec{k}s} = \sqrt{n} |n\rangle_{\vec{k}s}, \quad \text{and} \quad a_{\vec{k}s} |n\rangle_{\vec{k}s} = \sqrt{n} |n-1\rangle_{\vec{k}s}. \tag{2.83}$$

Equation (2.83) demonstrates that the creation operator adds one excitation (photon) to the EM field while the annihilation operator removes one excitation (photon) from the EM field. In particular, $a_{\vec{k}s}|0\rangle_{\vec{k}s} = 0$, so that no further photon can be removed from the ground state, which is thus also called the vacuum state. In addition, it can be shown that

$$N_{\vec{k}s} |n\rangle_{\vec{k}s} = n |n\rangle_{\vec{k}s}, \tag{2.84}$$

and

$$\hbar\omega_k \left(a^{\dagger}_{\vec{k}s} a_{\vec{k}s} + \frac{1}{2} \right) |n\rangle_{\vec{k}s} = \hbar\omega_k \left(n + \frac{1}{2} \right) |n\rangle_{\vec{k}s}, \tag{2.85}$$

with n being any nonnegative integer, so that the n-photon state has an energy $n\hbar\omega_k$ above the vacuum. The offset $\hbar\omega_k/2$, known as the ground (or vacuum) state energy of the field, is often neglected in the studies of quantum optics except when effects arising from the interaction with the vacuum are concerned, such as the

Casimir effect [13]. The set of states $\left\{|n\rangle_{\vec{k}s}\right\}$ forms a basis known as the Fock basis, with $|n\rangle_{\vec{k}s}$ called the Fock state or photon number state of the EM field at mode $\vec{k}s$.

Since the Fock states form a basis, any linear combination of them is also a quantum state of the EM field. In particular, the coherent state is defined as the eigenket of the annihilation operator:

$$a_{\vec{k}s}|\alpha\rangle_{\vec{k}s} = \alpha_{\vec{k}s}|\alpha\rangle_{\vec{k}s}, \tag{2.86}$$

in which

$$|\alpha\rangle_{\vec{k}s} = e^{-\frac{|\alpha|^2}{2}} \sum_{n=0}^{\infty} \frac{\alpha^n}{\sqrt{n!}} |n\rangle_{\vec{k}s}, \tag{2.87}$$

and α is any complex scalar. One should be cautious not to confuse $|\alpha\rangle_{\vec{k}s}$ with $|n\rangle_{\vec{k}s}$. By convention, the former refers to a coherent state in mode $\vec{k}s$ with the complex amplitude α while the latter refers to a Fock state with n photons in mode $\vec{k}s$.

The coherent state has many interesting and important properties [13]. The probability of the coherent state $|\alpha\rangle_{\vec{k}s}$ with n photons is

$$P_n = \left|\langle n|\alpha\rangle_{\vec{k}s}\right|^2 = e^{-|\alpha|^2} \frac{|\alpha|^{2n}}{n!}, \tag{2.88}$$

which is the Poisson distribution. The mean photon number of the coherent state is thus

$$\sum_{n=0}^{\infty} nP_n = {}_{\vec{k}s}\langle\alpha|N_{\vec{k}s}|\alpha\rangle_{\vec{k}s} = |\alpha|^2. \tag{2.89}$$

In fact, the ideal output of a single mode laser is a coherent state [14]. It should be noted that two different coherent states are not orthogonal, i.e., $\langle\beta|\alpha\rangle \neq 0$. It is also remarked that the coherent state with $\alpha = 0$ is identical to the vacuum state.

For the general multi-mode field, a quantum state can be expanded by the multi-mode Fock basis constructed by the tensor product: $\left\{|\{n\}\rangle \neq |n_{\vec{k}_1 s_1}\rangle|n_{\vec{k}_2 s_2}\rangle|n_{\vec{k}_3 s_3}\rangle\cdots\right\}$, in which $|n_{\vec{k}_j s_j}\rangle$ denotes the Fock state with $n_{\vec{k}_j s_j}$ photons in mode $\vec{k}_j s_j$. Explicitly, a multi-mode state can be written as

$$\sum_{n_{\vec{k}_1 s_1}, n_{\vec{k}_2 s_2}, n_{\vec{k}_3 s_3}, \ldots} C_{n_{\vec{k}_1 s_1}, n_{\vec{k}_2 s_2}, n_{\vec{k}_3 s_3}, \ldots} |n_{\vec{k}_1 s_1}\rangle|n_{\vec{k}_2 s_2}\rangle|n_{\vec{k}_3 s_3}\rangle\ldots, \tag{2.90}$$

where the coefficient $C_{n_{\vec{k}_1 s_1}, n_{\vec{k}_2 s_2}, n_{\vec{k}_3 s_3}, \ldots}$ manifests the quantum correlation among the different photon number components and modes. A corresponding multi-mode coherent state can also be written.

2.2.3 Representing Qubit Using Polarization States of a Photon

Consider the case of the EM field with a fixed wave vector $\vec{k}$. Suppressing the wave vector subscript, the electric field operator can be written as

$$\vec{E}(\vec{r},t) = E_k\left[(\vec{\epsilon}_1 a_1 + \vec{\epsilon}_2 a_2)e^{i\left(\vec{k}\cdot\vec{r}-\omega_k t\right)} - (\vec{\epsilon}_1 a_1 + \vec{\epsilon}_2 a_2)^{\dagger} e^{-i\left(\vec{k}\cdot\vec{r}-\omega_k t\right)}\right], \tag{2.91}$$

where $E_k = -i\sqrt{\hbar\omega_k/2\epsilon_0 V}$. For the two orthogonal polarization vectors expressed in terms of the horizontal (*H*) and vertical (*V*) polarizations:

$$\vec{\epsilon}_1 = \cos\frac{\theta}{2}\vec{x} + e^{i\varphi}\sin\frac{\theta}{2}\vec{y}, \quad \vec{\epsilon}_2 = -\sin\frac{\theta}{2}\vec{x} + e^{i\varphi}\cos\frac{\theta}{2}\vec{y}, \tag{2.92}$$

where $\theta \in [0,\pi]$ and $\varphi \in [0,2\pi]$, it can be shown that the creation operators read as

$$\begin{aligned} a_1^{\dagger}(\theta,\varphi) &= \cos\frac{\theta}{2}a_H^{\dagger} + e^{i\varphi}\sin\frac{\theta}{2}a_V^{\dagger}, \\ a_2^{\dagger}(\theta,\phi) &= -\sin\frac{\theta}{2}a_H^{\dagger} + e^{i\varphi}\cos\frac{\theta}{2}a_V^{\dagger}. \end{aligned} \tag{2.93}$$

The physical realization of a qubit by the polarization of a single photon is then given by

$$|\psi\rangle = \left(\alpha a_1^{\dagger} + \beta a_2^{\dagger}\right)|0\rangle_{\vec{k}}, \tag{2.94}$$

where we have retained the subscript $\vec{k}$ in the vacuum state $|0\rangle_{\vec{k}}$. We shall customarily use the horizontal and vertical polarizations to denote the computational basis $|0\rangle$ and $|1\rangle$ for the qubit introduced in Sect. 2.2.1. In this case, $\theta = 0$ and $\varphi = 0$, so that

$$|\psi\rangle = \left(\alpha a_H^{\dagger} + \beta a_V^{\dagger}\right)|0\rangle_{\vec{k}} = \alpha|0\rangle + \beta|1\rangle, \tag{2.95}$$

where $|0\rangle = a_H^{\dagger}|0\rangle_{\vec{k}}$ and $|1\rangle = a_V^{\dagger}|0\rangle_{\vec{k}}$, and α and β are defined in the same way as Eq. (2.1) with $|\alpha|^2 + |\beta|^2 = 1$. It should be emphasized that the label j in the qubit computational basis $\{|j\rangle, j = 0,1\}$, refers to the two events of a single photon in the *H* and *V* polarizations respectively, and the qubit state $|0\rangle$ should not be confused with the photon vacuum state $|0\rangle_{\vec{k}}$. It is also remarked that (θ,φ) defined in Eq. (2.93) are not exactly the same as the angles defined for the Bloch representation of a qubit in Eq. (2.17), even though these angles are related. This is explained in detail in the next section.

Finally, multiple qubits can be described by the creation operators by attributing each qubit to a particular wave vector mode $\vec{k}$. In this way, n qubits correspond to n distinguishable photons occupying n distinct wave vector modes.

2.2.4 Multi-photon Polarization States and Stokes Vector

The Fock state formalism allows us to describe multi-photon quantum states easily. For example, an n-photon quantum state can in principle be defined in a similar fashion to Eq. (2.94):

$$|\psi^{(n)}\rangle = \frac{1}{\sqrt{n!}}\left[\alpha\left(a_1^\dagger\right)^n + \beta\left(a_2^\dagger\right)^n\right]|0\rangle_{\vec{k}}. \tag{2.96}$$

This state is a highly nonclassical quantum state and is called the NOON state when $\alpha = \beta = 1/\sqrt{2}$, which finds useful applications in quantum metrology [15]. In general, an arbitrary quantum states can be written as $|\psi\rangle = f\left(a_1^\dagger, a_2^\dagger\right)|0\rangle_{\vec{k}}$, where $f(x_1, x_2)$ is an arbitrary function that guarantees $|\psi\rangle$ is normalizable. As an example, an n-photon state with polarization angles (θ, φ) defined according to Eq. (2.93) is given by

$$|n\rangle_{(\theta,\varphi)} = \frac{1}{\sqrt{n!}}\left[a_1^\dagger(\theta,\phi)\right]^n|0\rangle_{\vec{k}} = \frac{1}{\sqrt{n!}}\left(\cos\frac{\theta}{2}a_H^\dagger + e^{i\varphi}\sin\frac{\theta}{2}a_V^\dagger\right)^n|0\rangle_{\vec{k}}. \tag{2.97}$$

Correspondingly, a coherent state with the polarization angles (θ, φ) is

$$\begin{aligned}|\alpha\rangle_{(\theta,\varphi)} &= e^{-\frac{|\alpha|^2}{2}}\sum_{n=0}^{\infty}\frac{\alpha^n}{\sqrt{n!}}|n\rangle_{(\theta,\varphi)} = e^{-\frac{|\alpha|^2}{2}}\sum_{n=0}^{\infty}\frac{\alpha^n}{n!}\left(\cos\frac{\theta}{2}a_H^\dagger + e^{i\varphi}\sin\frac{\theta}{2}a_V^\dagger\right)^n|0\rangle_{\vec{k}} \\ &= e^{-\frac{|\alpha|^2}{2}}\sum_{n=0}^{\infty}\sum_{p=0}^{n}\frac{\alpha^n}{p!(n-p)!}\left(\cos\frac{\theta}{2}\right)^p\left(e^{i\varphi}\sin\frac{\theta}{2}\right)^{n-p}\left(a_H^\dagger\right)^p\left(a_V^\dagger\right)^{n-p}|0\rangle_{\vec{k}} \\ &= e^{-\frac{|\alpha|^2}{2}}\sum_{p=0}^{\infty}\sum_{q=0}^{\infty}\frac{\alpha^{p+q}}{p!q!}\left(\cos\frac{\theta}{2}\right)^p\left(e^{i\varphi}\sin\frac{\theta}{2}\right)^q\left(a_H^\dagger\right)^p\left(a_V^\dagger\right)^q|0\rangle_{\vec{k}} \\ &= \left|\alpha\cos\frac{\theta}{2}\right\rangle_H\left|\alpha e^{i\varphi}\sin\frac{\theta}{2}\right\rangle_V,\end{aligned} \tag{2.98}$$

where $|\alpha\rangle_H$ and $|\alpha\rangle_V$ are the coherent states with amplitude α in the polarization mode H and V respectively.

In classical optics, the polarization state of light can be described using the Jones vector and the Stokes vector representations. The Jones vector representation is obtained simply by specifying $\vec{\epsilon}_1$ and $\vec{\epsilon}_2$ in Eq. (2.91), which are usually taken to be horizontal and vertical polarizations:

$$\vec{\epsilon}_1 = \vec{x} = \begin{bmatrix} 1 \\ 0 \end{bmatrix}, \quad \text{and} \quad \vec{\epsilon}_2 = \vec{y} = \begin{bmatrix} 0 \\ 1 \end{bmatrix}. \tag{2.99}$$

In this way,

$$\vec{E}(\vec{r}, t) = E_k\left(\begin{bmatrix} a_H \\ a_V \end{bmatrix} e^{i\left(\vec{k}\cdot\vec{r}-\omega_k t\right)} - \begin{bmatrix} a_H \\ a_V \end{bmatrix}^{\dagger} e^{-i\left(\vec{k}\cdot\vec{r}-\omega_k t\right)}\right). \tag{2.100}$$

Hence the Jones vector is given by the operator

$$\vec{J} = \begin{bmatrix} a_H \\ a_V \end{bmatrix}, \tag{2.101}$$

which, for the coherent state, gives the expectation value

$$\langle\alpha|\vec{J}|\alpha\rangle_{(\theta,\varphi)} = \alpha \begin{bmatrix} \cos\frac{\theta}{2} \\ e^{i\varphi}\sin\frac{\theta}{2} \end{bmatrix}. \tag{2.102}$$

It should be noted that the expectation of $\vec{J}$ for the Fock state is zero, i.e., $\langle n|\vec{J}|n\rangle_{(\theta,\varphi)} = 0$, which is a manifestation of the quantumness of the photon number state [14].

On the other hand, the Stokes vector representation can be obtained through the Schwinger boson representation of SU(2) algebra:

$$\vec{S} = \begin{bmatrix} S_0 \\ S_1 \\ S_2 \\ S_3 \end{bmatrix} = \begin{bmatrix} a_H^\dagger a_H + a_V^\dagger a_V \\ a_H^\dagger a_H - a_V^\dagger a_V \\ a_H^\dagger a_V + a_V^\dagger a_H \\ -i\left(a_H^\dagger a_V - a_V^\dagger a_H\right) \end{bmatrix} = \begin{bmatrix} a_H^\dagger a_H + a_V^\dagger a_V \\ a_H^\dagger a_H - a_V^\dagger a_V \\ a_D^\dagger a_D - a_A^\dagger a_A \\ a_L^\dagger a_L - a_R^\dagger a_R \end{bmatrix}, \tag{2.103}$$

which results in the expectation for the coherent state

$$\langle\alpha|\vec{S}|\alpha\rangle_{(\theta,\varphi)} = |\alpha|^2 \begin{bmatrix} 1 \\ \cos\theta \\ \sin\theta\cos\varphi \\ \sin\theta\sin\varphi \end{bmatrix}. \tag{2.104}$$

Here $a_D^\dagger = \left(a_H^\dagger + a_V^\dagger\right)/\sqrt{2}$, $a_A^\dagger = \left(a_H^\dagger - a_V^\dagger\right)/\sqrt{2}$, $a_L^\dagger = \left(a_H^\dagger + ia_V^\dagger\right)/\sqrt{2}$, and $a_R^\dagger = \left(a_H^\dagger - ia_V^\dagger\right)/\sqrt{2}$. Therefore, it can be seen that the expectation value of the

Stokes vector with respect to the coherent state gives the same result as that in classical optics. The normalized Stokes vector

$$\vec{s} = \begin{bmatrix} s_1 \\ s_2 \\ s_3 \end{bmatrix} = \frac{1}{\langle\alpha|S_0|\alpha\rangle} \begin{bmatrix} \langle\alpha|S_1|\alpha\rangle \\ \langle\alpha|S_2|\alpha\rangle \\ \langle\alpha|S_3|\alpha\rangle \end{bmatrix}_{(\theta,\varphi)} = \begin{bmatrix} \cos\theta \\ \sin\theta\,\cos\varphi \\ \sin\theta\,\sin\varphi \end{bmatrix} \tag{2.105}$$

can be visualized graphically through the Poincaré sphere as depicted in Fig. 2.2.

Some comments are in order. First, the angles θ and φ here are defined with respect to the horizontal (x-axis) and the diagonal (y-axis) polarizations, whereas they are defined in Sect. 2.2.1 for the Bloch sphere with respect to the $|0\rangle$ (z-axis) and $|+\rangle$ (x-axis) qubits. The normalized Stokes vector can be reparametrized by the usual spherical coordinates of the Poincaré sphere (see Fig. 2.2) by writing the creation operator as

$$\begin{aligned} a^{\dagger}(\chi,\psi) &= \cos\left(\frac{\pi}{4}-\chi\right)a_L^{\dagger} + e^{2i\psi}\sin\left(\frac{\pi}{4}-\chi\right)a_R^{\dagger} \\ &= e^{i\psi}\left[(\cos\chi\,\cos\psi - i\,\sin\chi\,\sin\psi)a_H^{\dagger} + (\cos\chi\,\sin\psi + i\,\sin\chi\,\cos\psi)a_V^{\dagger}\right], \end{aligned} \tag{2.106}$$

so that

$$\vec{s} = \begin{bmatrix} s_1 \\ s_2 \\ s_3 \end{bmatrix} = \begin{bmatrix} \cos 2\chi\,\cos 2\psi \\ \cos 2\chi\,\sin 2\psi \\ \sin 2\chi \end{bmatrix}. \tag{2.107}$$

Second, the expectation value of the Stokes vector is well-defined for any quantum state even though the corresponding expectation value of the Jones vector may not exist (such as the Fock states). This can be seen through the polarization rotation explained in the next section.

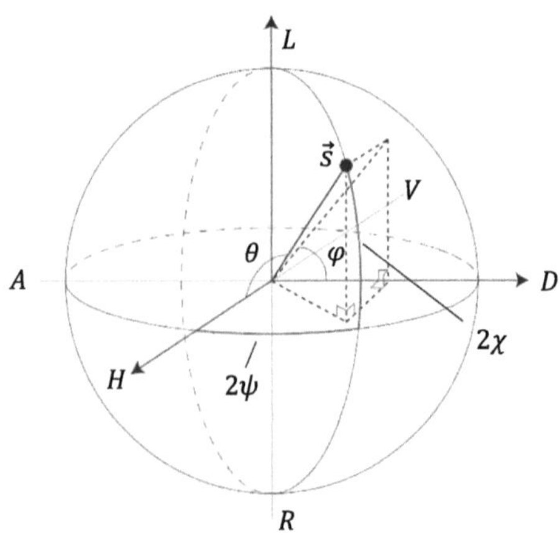

Fig. 2.2 Poincaré sphere representation of the expectation value of the normalized Stokes vector with respect to the coherent state. The coordinates (θ, φ) corresponds to the polarization of the coherent state defined by Eq. (2.93) whereas the coordinates (χ, ψ) corresponds to Eq. (2.106)

2.2.5 Polarization Rotation and Mueller Matrices for Multi-photon States

The creation operations in Eq. (2.93) can be interpreted as the horizontal and vertical mode operators rotated through an angle θ and retarded by a phase shift φ:

$$a_1^\dagger(\theta,\varphi) = U_{(\theta,\varphi)} a_H^\dagger U_{(\theta,\varphi)}^\dagger, \quad a_2^\dagger(\theta,\varphi) = U_{(\theta,\varphi)} a_V^\dagger U_{(\theta,\varphi)}^\dagger, \tag{2.108}$$

where $U_{(\theta,\varphi)} = U_{\text{shft}}(\varphi) U_{\text{rot}}(\theta)$ with

$$U_{\text{rot}}(\theta) = e^{\frac{\theta}{2}\left(a_V^\dagger a_H - a_H^\dagger a_V\right)}, \quad U_{\text{shft}}(\varphi) = e^{i\varphi a_V^\dagger a_V}. \tag{2.109}$$

Here $U_{\text{rot}}(\theta)$ is the rotation operator for the modes a_H and a_V, whereas $U_{\text{shft}}(\varphi)$ accounts for the phase shift φ of the vertical polarization with respect to the horizontal polarization. This can be seen more explicitly by applying $U_{(\theta,\varphi)}$ to the Jones vector:

$$U_{(\theta,\varphi)}^\dagger \vec{J} U_{(\theta,\varphi)} = U_{\text{rot}}(-\theta) U_{\text{shft}}(-\varphi) \vec{J} U_{\text{shft}}^\dagger(-\varphi) U_{\text{rot}}^\dagger(-\theta) = Q_{\text{shft}}(\varphi) Q_{\text{rot}}(\theta) \vec{J}, \tag{2.110}$$

where

$$Q_{\text{rot}}(\theta) = \begin{bmatrix} \cos\frac{\theta}{2} & -\sin\frac{\theta}{2} \\ \sin\frac{\theta}{2} & \cos\frac{\theta}{2} \end{bmatrix}, \quad Q_{\text{shft}}(\varphi) = \begin{bmatrix} 1 & 0 \\ 0 & e^{i\varphi} \end{bmatrix}. \tag{2.111}$$

This shows a polarization rotation by φ followed by a phase retardation. Equation (2.110) is obtained using Eq. (2.108) with the realization that $U_{\text{rot}}^\dagger(\theta) = U_{\text{rot}}(-\theta)$ and $U_{\text{shft}}^\dagger(\varphi) = U_{\text{shft}}(-\varphi)$. Likewise, it can be shown that the Stokes vector follows the relation below:

$$U_{(\theta,\varphi)}^\dagger \vec{S} U_{(\theta,\varphi)} = M_{\text{shft}}(\varphi) M_{\text{rot}}(\theta) \vec{S}, \tag{2.112}$$

where

$$M_{\text{rot}}(\theta) = \begin{bmatrix} 1 & 0 & 0 & 0 \\ 0 & \cos\theta & -\sin\theta & 0 \\ 0 & \sin\theta & \cos\theta & 0 \\ 0 & 0 & 0 & 1 \end{bmatrix}, \tag{2.113a}$$

$$M_{\text{shft}}(\varphi) = \begin{bmatrix} 1 & 0 & 0 & 0 \\ 0 & 1 & 0 & 0 \\ 0 & 0 & \cos\varphi & -\sin\varphi \\ 0 & 0 & \sin\varphi & \cos\varphi \end{bmatrix}. \tag{2.113b}$$

$M_{\text{rot}}(\theta)$ and $M_{\text{shft}}(\varphi)$ are the Mueller matrices of (active) rotation at the angle $\theta/2$ and phase retardation of the vertical polarization with the phase φ with respect to the horizontal polarization [16].

Consider a quantum state with a fixed polarization with angles (θ, φ). The quantum state can generally be written as

$$|\psi\rangle_{(\theta,\varphi)} = f\left(a_1^\dagger(\theta,\varphi)\right)|0\rangle_{\vec{k}}, \tag{2.114}$$

for some function $f(x)$. According to Eq. (2.108) and using the Taylor series expansion of $f(x)$, we can rewrite (2.114) as

$$\begin{aligned} |\psi\rangle_{(\theta,\varphi)} &= \sum_{m=0}^{\infty} \frac{f^{(m)}(0)}{m!}\left[a_1^\dagger(\theta,\varphi)\right]^m |0\rangle_{\vec{k}} = \sum_{m=0}^{\infty} \frac{f^{(m)}(0)}{m!}\left[U_{(\theta,\varphi)} a_H^\dagger U_{(\theta,\varphi)}^\dagger\right]^m |0\rangle_{\vec{k}} \\ &= U_{(\theta,\varphi)} \sum_{m=0}^{\infty} \frac{f^{(m)}(0)}{m!}\left(a_H^\dagger\right)^m U_{(\theta,\varphi)}^\dagger |0\rangle_{\vec{k}} = U_{(\theta,\varphi)} f\left(a_H^\dagger\right)|0\rangle_{\vec{k}} \\ &= U_{(\theta,\varphi)} |\psi\rangle_{(\theta=0,\varphi=0)}. \end{aligned} \tag{2.115}$$

Note that we have used $U_{(\theta,\varphi)}^\dagger |0\rangle_{\vec{k}} = |0\rangle_{\vec{k}}$. Therefore,

$$\begin{aligned} \langle\psi|\vec{S}|\psi\rangle_{(\theta,\varphi)} &= \langle\psi|U_{(\theta,\varphi)}^\dagger \vec{S} U_{(\theta,\varphi)}|\psi\rangle_{(\theta=0,\varphi=0)} = M_{\text{shft}}(\varphi)M_{\text{rot}}(\theta)\langle\psi|\vec{S}|\psi\rangle_{(\theta=0,\varphi=0)} \\ &= M_{\text{shft}}(\varphi)M_{\text{rot}}(\theta)\begin{bmatrix} \langle N\rangle \\ \langle N\rangle \\ 0 \\ 0 \end{bmatrix} = \langle N\rangle \begin{bmatrix} 1 \\ \cos\theta \\ \sin\theta\cos\varphi \\ \sin\theta\sin\varphi \end{bmatrix}, \end{aligned} \tag{2.116}$$

where $\langle N\rangle = \langle\psi|a_H^\dagger a_H|\psi\rangle_{(\theta=0,\varphi=0)}$ is the mean photon number of the state $|\psi\rangle_{(\theta,\varphi)}$. This demonstrates that the expectation value of the Stokes vector is defined for an arbitrary quantum state with a well-defined polarization.

Equation (2.112) manifests that the quantum Stokes vector $\vec{S}$ can be operated as in classical optics through the usual (lossless) Mueller matrices $M_{\text{shft}}(\varphi)$ and $M_{\text{rot}}(\theta)$ irrespective of the quantum states of the EM field. The Mueller matrix for a linear retarder with retardance δ and fast axis oriented an angle θ from the x axis is given by [16]

$$M_{\delta}(\theta) = M_{\text{rot}}(2\theta)M_{\text{shft}}(\varphi)M_{\text{rot}}(-2\theta)$$
$$= \begin{bmatrix} 1 & 0 & 0 & 0 \\ 0 & \cos^2 2\theta + \cos\delta\,\sin^2 2\theta & (1-\cos\delta)\cos 2\theta\,\sin 2\theta & \sin\delta\,\sin 2\theta \\ 0 & (1-\cos\delta)\cos 2\theta\,\sin 2\theta & \sin^2 2\theta + \cos\delta\,\cos^2 2\theta & -\sin\delta\,\cos 2\theta \\ 0 & -\sin\delta\,\sin 2\theta & \sin\delta\,\cos 2\theta & -1 \end{bmatrix}. \tag{2.117}$$

For example, the Mueller matrix of a half wave plate is

$$M_{\text{HWP}}(\theta) = M_{\delta=\pi}(\theta) = \begin{bmatrix} 1 & 0 & 0 & 0 \\ 0 & \cos 4\theta & \sin 4\theta & 0 \\ 0 & \sin 4\theta & -\cos 4\theta & 0 \\ 0 & 0 & 0 & -1 \end{bmatrix}. \tag{2.118}$$

2.3 Summary

This chapter has introduced the basic concept of a qubit and its quantum states. Then the photons are defined rigorously as the excitations of the quantized electromagnetic field and the representation of a qubit by the photon polarization state is given. Finally, it has discussed polarization rotation for a stream of photons and its representation in terms of Mueller Matrices.

References

1. Nielsen, M. A., & Chuang, I. L. (2000). *Quantum computation and quantum information*. Cambridge: Cambridge University Press.
2. Dirac, P. A. M. (1981). *The principles of quantum mechanics*. Oxford: Oxford University Press.
3. Dieks, D. (1982). Communication by EPR devices. *Physics Letters A, 92*(6), 271–272.
4. Wootters, W. K., & Zurek, W. H. (1982). A single quantum cannot be cloned. *Nature, 299* (5886), 802–803.
5. Bennett, C. H., & Brassard, G. (2014). Quantum cryptography: Public key distribution and coin tossing. *Theoretical Computer Science, 560,* 7–11.
6. Barnum, H., et al. (1996). Noncommuting mixed states cannot be broadcast. *Physical Review Letters, 76*(15), 2818.
7. Bužek, V., & Hillery, M. (1996). Quantum copying: Beyond the no-cloning theorem. *Physical Review A, 54*(3), 1844.
8. Gisin, N., & Massar, S. (1997). Optimal quantum cloning machines. *Physical Review Letters, 79*(11), 2153.

9. Werner, R. F. (1998). Optimal cloning of pure states. *Physical Review A, 58*(3), 1827.
10. Scarani, V., et al. (2005). Quantum cloning. *Reviews of Modern Physics, 77*(4), 1225–1256.
11. Lemm, M., & Wilde, M. M. (2017). Information-theoretic limitations on approximate quantum cloning and broadcasting. *Physical Review A, 96*(1), 012304.
12. Jackson, J. D. (1999). *Classical electrodynamics* (3rd ed.). Hoboken: Wiley.
13. Walls, D. F., & Milburn, G. J. (2007). *Quantum optics*. Berlin: Springer Science & Business Media.
14. Scully, M. O., & Zubairy, M. S. (1999). *Quantum optics*. AAPT.
15. Giovannetti, V., Lloyd, S., & Maccone, L. (2011). Advances in quantum metrology. *Nature Photonics, 5*(4), 222.
16. Bass, M. (2000). *Handbook of optics*. New York: McGraw-Hill.

Chapter 3
Quantum Key Distribution

Conventional implementations of cryptography are only computationally secure. The security of quantum cryptography (QC), on the other hand, is based on the inherent uncertainty in quantum phenomena at the physical layer of a communication system. Quantum key distribution (QKD), the most well-known and developed application of QC to establish a shared key between two parties, is provably unbreakable even with unlimited computational power. The intent of this chapter is to provide a brief survey of several common QKD protocols including the BB84, B92, SARG04, COW, decoy-state BB84, and E91 protocols.

3.1 Introduction

The appeal of QKD is that it is a provably secure key distribution approach which is not based on a one-way mathematical function. One-way mathematical functions are breakable with finite computational resources [1]. The security of QKD exploits the physical laws of quantum mechanics that can detect the presence of an eavesdropper Eve, and limit the amount of information that can be leaked to her.

A QKD protocol usually relies on a *quantum channel* and a *classical channel*. The quantum channel is a physical communication channel for the transmission of information. Photons representing data (usually random numbers) encoded in a chosen physical degree of freedom of the photons (e.g., polarization) are transmitted through the quantum channel. The transmission can be from one user Alice to another user Bob (one-way protocol) or between them (two-way protocol). In this process, Eve can tap onto the quantum channel, but her capability is limited by the laws of quantum physics. On the other hand, the classical channel is for the communication to post-process the information shared between Alice and Bob to establish a final secure key. This post-processing includes *sifting* (to retain the useful information), *error correction*, and *privacy amplification*. The sifting step depends on the specific protocol used. Error correction is needed because the shared

P. K. Verma et al., *Multi-photon Quantum Secure Communication*,
Signals and Communication Technology,
https://doi.org/10.1007/978-981-10-8618-2_3

information through the quantum channel can be erroneous due to the effects of noise in the quantum channel or the eavesdropping by Eve. The reason why intrusion induces error will be discussed in this Chapter. In fact, it is exactly the errors imposed by Eve that give the quantitative security of QKD.

Privacy amplification is a step used to make the information leaked to Eve negligible. In fact, all errors caused in the transmission are all attributed to Eve. It should be noted that the classical channel is public so that the legitimate users need to ensure that information shared through the classical channel does not lead to the additional leakage of information shared through the quantum channel. The public classical channel also needs to be authenticated to avoid any man-in-the-middle (MIM) attack by the eavesdropper.

3.2 Single Photon-Based QKD Protocols

The basic idea of QKD is that when information is encoded as a qubit in a single photon, the eavesdropper cannot gain any information without disturbing the state of the qubit. This is based on the fact that a single photon cannot be split into more than one entity. Then by virtue of the no-cloning theorem presented in Chap. 2, the eavesdropper cannot exactly copy the state of the qubit. In addition, the process of distinguishing between two non-orthogonal quantum states inevitably disturbs either of the quantum states. This can be seen through the argument that is similar to the no-cloning theorem below [2].

Suppose the two quantum states $|\psi\rangle$ and $|\phi\rangle$ are non-orthogonal and normalized, and the eavesdropper wishes to know which is which. To gain information, what the eavesdropper does is to apply a unitary operation U to interact the quantum state (either $|\psi\rangle$ and $|\phi\rangle$) to her system $|\eta\rangle$. If this process does not disturb the given quantum state, we can write

$$|\psi\rangle|\eta\rangle \rightarrow |\psi\rangle|\eta'\rangle = U|\psi\rangle|\eta\rangle, \tag{3.1}$$

$$|\phi\rangle|\eta\rangle \rightarrow |\phi\rangle|\eta''\rangle = U|\phi\rangle|\eta\rangle. \tag{3.2}$$

By taking the scalar product using Eqs. (3.1) and (3.2), we obtain

$$\langle\psi|\phi\rangle\langle\eta'|\eta''\rangle = \langle\psi|\phi\rangle\langle\eta|\eta\rangle, \tag{3.3}$$

giving $\langle\eta'|\eta''\rangle = \langle\eta|\eta\rangle = 1$. Thus $|\eta'\rangle$ and $|\eta''\rangle$ are identical, implying that the eavesdropper cannot distinguish between $|\psi\rangle$ and $|\phi\rangle$ by checking the state she owns using the operation applied. Then, by contradiction, in order to distinguish between $|\psi\rangle$ and $|\phi\rangle$, the eavesdropper's operation has to disturb the input state, which then possibly causes errors in the legitimate users' measurements. In the following, two common single-photon based QKD protocols are described.

3.2.1 The BB84 Protocol

Proposed by Bennett and Brassard in 1984 [3], the BB84 protocol is arguably the first application of quantum physics in information and communication theory that gave rise to the explosive investigation of quantum communication and cryptography. In a typical scenario of BB84 protocol, the legitimate users, Alice and Bob, establish a secret random binary string, which will be used later as a secret cryptographic key. The BB84 protocol proceeds as follow:

1. Alice prepares two strings $x_A = x_{A1}x_{A2}\cdots x_{AN}$ and $y_A = y_{A1}y_{A2}\cdots y_{AN}$ of random classical bits.
2. Alice uses x_A and y_A to form a quantum state as a tensor product of N qubits

$$|\Psi\rangle = \bigotimes_{k=1}^{N} |\psi_{x_{Ak}y_{Ak}}\rangle, \tag{3.4}$$

where $|\psi_{x_k y_k}\rangle$ is given by, depending on the values of x_{Ak} and y_{Ak},

$$|\psi_{00}\rangle = |0\rangle, \tag{3.5a}$$

$$|\psi_{10}\rangle = |1\rangle, \tag{3.5b}$$

$$|\psi_{01}\rangle = |+\rangle = \frac{|0\rangle + |1\rangle}{\sqrt{2}}, \tag{3.5c}$$

$$|\psi_{11}\rangle = |-\rangle = \frac{|0\rangle - |1\rangle}{\sqrt{2}}. \tag{3.5d}$$

Essentially, the string y_A is used to choose between the bases of the operators Z $(\{|0\rangle, |1\rangle\})$ and X $(\{|+\rangle, |-\rangle\})$, and the string x_A is used to fix the particular qubit in the basis.

3. Alice sends the sequence of qubits to Bob through the quantum channel.
4. Bob receives the qubits and makes measurement on the qubits using either the basis X or Z for each qubit according to a randomly generated sequence y_B. The measurement x_B results in a bit value 0 (1) if the measurement corresponds to the positive (negative) eigenvalue of X or Z. Bob then announces that he received the qubits through the public classical channel.
5. It should be noted that, for a particular qubit, if Bob's measurement basis is the same as Alice's preparation basis, Bob's measured bit value in x_B will be the same as Alice's bit value in x_A. If the bases are different, since the basis sets X and Z are mutually unbiased, i.e., the scalar product of a basis ket in X and a basis ket in Z gives a probability of 0.5, Bob's resulting measured bit value will

have a probability of 0.5 being correct (or incorrect). To determine whether they use the same set of bases, Alice announces the string y_A in the classical channel.

6. Alice and Bob discuss through the classical channel to discard those bits in x_A and x_B that the preparation basis by Alice and the measurement basis by Bob are different.
7. At this point, for each of the remaining bits in x_A (called the sifted key x'_A), Alice's preparation basis and Bob's measurement basis are the same. If no errors are incurred in the qubits during the transmission in the quantum channel, Bob's measured states for those qubits should give rise to bit values x'_B matching the bits in x'_A, i.e., $x'_A = x'_B$. However, it is very likely that errors occurred in the qubits during the transmission. Alice thus selects a subset of x'_A and tells Bob which bits are selected through the classical channel.
8. Alice and Bob check the values of the selected bits through the classical channel. If the error in those bits (called the quantum bit error rate, QBER) is higher than a threshold, the protocol is aborted. If not, they proceed to the next step. The threshold is determined by the security analysis that makes sure privacy amplification can be carried out.
9. Alice and Bob carry out error correction and privacy amplification through the classical channel.

For the case of no error, the Steps 1–7 are illustrated through the example in Table 3.1.

Table 3.1 Example of the working of the BB84 protocol

Step	Alice	Transmission									Bob
1	Bit sequence x_A	1	1	0	0	1	0	1	0	1	
	Bit sequence y_A	1	0	0	0	0	1	1	1	1	
2	Preparation basis	×	+	+	+	+	×	×	×	×	
	Qubit sequence	↘	↕	↔	↔	↕	↗	↘	↗	↘	
4		1	1	0	1	0	0	1	1	0	Bit sequence y_B
		×	×	+	×	+	+	×	×	+	Measurement basis
		1	1	0	1	1	1	1	0	1	Measured bit values x_B
6	Basis matching ($y_A = y_B$?)	✓	✗	✓	✗	✓	✗	✓	✓	✗	Basis matching ($y_A = y_B$?)
7	Sifted key x'_A	1	–	0	–	1	–	1	0	–	Sifted key $x'_B = x'_A$

In Step 8, the error threshold is determined by the condition that error correction and privacy amplification can lead to almost zero error in Bob's final key and negligible information leaked to Eve. This can be captured by the conditions that [4]

1. For a given joint probability $P(A, B, E)$ where A, B and E denote Alice, Bob and Eve, Alice and Bob can establish a secret key using classical error correction and privacy amplification if and only if the mutual information of Alice and Bob is greater than the mutual information of Eve with either Alice or Bob. That is,

$$I(A:B) \geq I(A:E), \quad \text{or} \quad I(A:B) \geq I(B:E), \tag{3.6}$$

where $I(A:B) \equiv H(A) - H(A|B)$ denotes the mutual information and H is the Shannon entropy [2].

2. The sum of Eve's and Bob's information per qubit is less than or equal to 1, that is, for n qubits,

$$I(A:B) + I(A:E) \leq n. \tag{3.7}$$

Combing Eqs. (3.6) and (3.7), we find a secret key is achievable if $I(A:B) \geq n/2$. Noting that $H(A) = 1$ and $I(A:B) = n[1 - h(Q)]$, where $h(x) = -x\log_2 x - (1-x)\log_2(1-x)$ is the binary entropy function and Q is the QBER, we obtain the condition for a secret key is

$$1 - 2h(Q) \geq 0, \tag{3.8}$$

or $Q \leq 11\%$. Therefore, the threshold in Step 8 is given by Bob's QBER $\leq 11\%$, the threshold that is commonly quoted [5]. In fact, Eq. (3.8) is related to the asymptotic secret key generation rate of BB84 [4–6]

$$K = \max(S[1 - 2h(Q)], 0), \tag{3.9}$$

where S is the gain (the ratio of the number of Bob's detection events using the same basis as Alice to Alice's number of emitted signals) of the protocol. It is remarked that although we use simple intuitions to arrive at Eq. (3.8), the same result can be derived using a more detailed analysis by treating the processes of error correction and privacy amplification explicitly [5, 6].

3.2.2 The B92 Protocol

As discussed earlier, the basis of QKD is to make use of non-orthogonal states to encode the information. The BB84 protocol accomplishes this by utilizing the eigenkets of two mutually unbiased basis X and Z. The total number of different qubit states that are transmitted is thus four. On the other hand, the B92 protocol, named after Bennett [7], makes use of only two of the four states. The two transmitted qubit states are

$$|\psi_0\rangle = |0\rangle, \tag{3.10a}$$

$$|\psi_1\rangle = |+\rangle = \frac{|0\rangle + |1\rangle}{\sqrt{2}}. \tag{3.10b}$$

There are slight modifications to the BB84 protocol. The B92 protocol proceeds as follow:

1. Alice prepares a string $x = x_1, x_2, \ldots, x_N$ of random classical bits.
2. Alice uses x to form a quantum state as a tensor product of N qubits

$$|\Psi\rangle = \bigotimes_{k=1}^{N} |\psi_{x_k}\rangle, \tag{3.11}$$

   where $|\psi_{x_k}\rangle$, $x_k = 0, 1$, is given by Eq. (3.10).
3. Alice sends the sequence of qubits to Bob through the quantum channel.
4. Bob prepares a string $y = y_1, y_2, \ldots, y_N$ of random classical bits. Bob receives the qubits and makes measurements on the qubits using the measurement operators

$$P_0 \equiv 1 - |\psi_1\rangle\langle\psi_1| = |-\rangle\langle-|, \quad \text{if } y_k = 0, \tag{3.12a}$$

$$P_1 \equiv 1 - |\psi_0\rangle\langle\psi_0| = |1\rangle\langle1|, \quad \text{if } y_k = 1, \tag{3.12b}$$

   for each qubit. It should be noted that the projections of the measurements on the states give

$$P_0|\psi_0\rangle = \frac{1}{\sqrt{2}}|-\rangle, \tag{3.13a}$$

$$P_0|\psi_1\rangle = 0, \tag{3.13b}$$

$$P_1|\psi_0\rangle = 0, \tag{3.13c}$$

$$P_1|\psi_1\rangle = \frac{1}{\sqrt{2}}|1\rangle. \tag{3.13d}$$

Therefore, the measurement results are non-empty (nonzero measurement probability) only when $x_k = y_k$. Bob then can construct a sequence of bits z denoting whether the measurement is empty or non-empty, denoted by 0 and 1, respectively. Bob then announces the bit sequence z through the public classical channel while keeping y secret.

5. Alice and Bob discuss through the classical channel to keep only those pairs of bits $\{x_k, y_k\}$ with $z_k = 1$. The sifted key x' for Alice and Bob is the pairs of the retained bits $\{x_k, y_k\}$, which are perfectly correlated if they are not contaminated by noise.
6. The remaining steps are the same as Steps 7–9 of BB84.

For the case of no error, the Steps 1–5 of B92 are illustrated through the example in Table 3.2.

3.3 Use of Weak Coherent States in QKD

The single photon-based QKD protocols are powerful schemes that provide physical security for the distribution of a secret key between the users Alice and Bob by limiting the eavesdropper Eve gaining information on the key without introducing errors in the correlations between Alice and Bob. However, genuine single photon sources are still an active area of research, especially at the telecom wavelengths [8–10]. All existing commercial QKD systems nevertheless rely on weak coherent pulse.

The use of coherent states with a small mean intensity implies that most of the time there are no photons (probability is $e^{-\mu} \approx 90.5\%$ for $\mu = 0.1$). This severely limits the transmission data rate. Worse still, there are chances that the pulses

Table 3.2 Example of the working of the B92 protocol

Step	Alice	Transmission									Bob
1	Bit sequence $\boldsymbol{x}$ for state selection	1	1	0	0	1	0	1	0	1	
2	Qubit sequence	↗	↗	↔	↔	↗	↔	↗	↔	↗	
4		1	0	0	0	0	1	1	1	1	Bit sequence $\boldsymbol{y}$ for basis selection
		∣	╲	╲	╲	╲	∣	∣	∣	∣	Measurement basis
		1	0	1	0	0	0	1	0	0	Measurement result $\boldsymbol{z}$
5	Measurement nonempty?	✓	✗	✓	✗	✗	✗	✓	✗	✗	Measurement nonempty?
	Sifted key $\boldsymbol{x}'$	1	–	0	–	–	–	1	–	–	Sifted key $\boldsymbol{x}'$

contain more than one photon. For example, when the mean intensity is 0.1, the probability of getting more than one photon is $1-(1+\mu)e^{-\mu} \approx 0.47\%$. This gives rise to the potential photon-number-splitting (PNS) attack [11].

3.3.1 Photon-Number-Splitting Attack

The PNS attack can be launched by Eve, who is assumed to be all-powerful except being limited by the laws of physics, using a photon number quantum non-demolition (QND) measurement [12, 13]. The QND measurement in principle enables her to know the number of photons in a light pulse without disturbing the qubit state of the photons. She can intercept all light pulses from Alice, block all single photon pulses, store one of the photons from multiphoton pulses in a quantum memory [14, 15], and send the rest of the photons from the multiphoton pulses to Bob. From Bob's perspective, the loss of the single photon pulses as blocked by Eve effectively is no different from the loss due to the imperfect quantum channel. For Eve, with the use of the quantum memory, she can delay her measurements on her photons until Bob announces his measurement basis for BB84. Even without quantum memory, Eve can still perform optimal unambiguous state discrimination on her qubits [16]. In this way, she can gain information without being caught. Similar strategies can be proposed to adapt to the different QKD protocols.

More concretely, when coherent states are used, the probability distribution of obtaining n photons follows Poisson distribution (when the phase information of the coherent states is hidden by Alice through phase randomization):

$$p_n = \frac{e^{-\mu}\mu^n}{n!}, \tag{3.14}$$

where μ is the mean photon number of the pulses. For the implementation of the BB84 protocol with weak pulses, Bob's raw detection rate is, in the absence of Eve, given by

$$R_{\text{Bob}}(\delta) = \sum_{n\geq 1}^{\infty} p_n[1-(1-\eta_{\text{det}}\eta_\delta)^n] = 1 - e^{-\eta_{\text{det}}\eta_\delta\mu} \approx \eta_{\text{det}}\eta_\delta\mu, \tag{3.15}$$

where η_{det} is Bob's device transmittance and detector efficiency and η_δ is the transmission coefficient of the channel. For optical fiber of length ℓ,

$$\eta_\delta = 10^{-\delta/10}, \quad \delta = \alpha\ell \text{ [dB]}. \tag{3.16}$$

For optical fiber at the telecom wavelength, the typical value for α is 0.25 dB/km.

To launch PNS attack using quantum memory without being noticed, Eve needs to constrain the attack so that Bob's raw detection rate (3.15) is not modified. To obtain complete information, Eve blocks all single photon pulses and replaces the optical fiber with a channel that has zero attenuation. Then Bob's maximum data rate in this case is given by

$$R_{\mathrm{Bob}}^{\mathrm{PNS}} = \sum_{n\geq 2}^{\infty} p_n[1-(1-\eta_{\mathrm{det}})^n] = 1 - e^{-\eta_{\mathrm{det}}\mu} - \eta_{\mathrm{det}}\mu e^{-\mu} \approx \eta_{\mathrm{det}} p_2, \tag{3.17}$$

As a result, the attenuation in the fiber has to be larger than the critical value δ_c defined by

$$R_{\mathrm{Bob}}(\delta_c) = R_{\mathrm{Bob}}^{\mathrm{PNS}}, \tag{3.18}$$

giving

$$\delta_c \approx 10 \log_{10} \frac{2e^{\mu}}{\mu}. \tag{3.19}$$

For pulse mean intensity $\mu = 0.1$, we get the critical attenuation $\delta_{\mathrm{BB84}} \approx 13$dB or fiber length $\ell_{\mathrm{BB84}} \approx 50$ km. For a shorter distance, Eve can no longer obtain full information without changing Bob's raw detection rate, though she can get partial information. In that case, Alice and Bob can use a privacy amplification scheme to obtain a secret but shorter key about which Eve has negligible information.

On the other hand, Eve can launch PNS attacks and gain full information without the use of quantum memory. This can be done by blocking both single-photon and two-photon pulses, and using those pulses with more than two photons together with unambiguous state discrimination. It was then shown that Eve can obtain conclusive results about the state of the pulses with probability 1/2 [17]. Then she can send a new photon prepared according to the conclusive result to Bob. Such an attack is also called the intercept-resend with unambiguous discrimination (IRUD) attack [17].

To obtain the secure key rate with coherent states, we can regard the potential PNS attacks for BB84 as essentially giving Eve the ability to tag the pulses that reveal the bases used in the preparation of the states. Accordingly, the secret key rate in Eq. (3.9) is modified as [4, 6, 18]

$$K_{\mathrm{coh}} = \max\left(S_{\mu}\left\{(1-\Delta)\left[1-h\left(\frac{Q_{\mu}}{1-\Delta}\right)\right] - h(Q_{\mu})\right\}, 0\right), \tag{3.20}$$

where S_μ is the gain (the ratio of the number of Bob's detection events using the same basis as Alice to Alice's number of emitted signals) of the protocol, Q_μ is the QBER (the error rate of Bob's detection events for the case of Alice and Bob using the same basis), and Δ is the fraction of tagged signals. Both S_μ and Q_μ are quantities measurable in the protocol, whereas Δ has to be estimated. In the worst situation where all the multiphoton events emitted by Alice are tagged by Eve and are received by Bob, the tag ratio can be estimated by

$$\Delta = \frac{p_{\text{multi}}}{S_\mu} \sim \frac{R_{\text{Bob}}^{\text{PNS}}}{R_{\text{Bob}}(\delta_c)}, \tag{3.21}$$

where p_{multi} is the probability of Alice emitting multi-photon pulses. Therefore, since $\Delta < 1$, under such a worst case scenario, the mean intensity μ has to be of order $O(\eta_\delta)$.

The gain S_μ is related to Bob's counting rates (or called yield) Y_n, which is defined as the conditional probability that Bob detects a signal given an n-photon signal emitted by Alice:

$$S_\mu = \sum_{n=0}^{\infty} Y_n p_n, \tag{3.22}$$

where p_n is given by Eq. (3.14) and

$$Y_n = Y_0 + \eta_n - Y_0\eta_n \approx Y_0 + \eta_n, \tag{3.23}$$

with $Y_0 \sim 10^{-5}$ being the background rate including detector dark count and stray light, and

$$\eta_n = 1 - (1 - \eta_{\text{det}}\eta_\delta)^n. \tag{3.24}$$

Using Eqs. (3.14), (3.22), (3.23) and (3.24), we obtain

$$S_\mu = Y_0 + 1 - e^{-\eta_{\text{det}}\eta_\delta\mu}. \tag{3.25}$$

With $Y_0 \ll \eta_\delta$ and $\mu \sim O(\eta_\delta)$, we find $S_\mu \sim O(\eta_\delta^2)$ and hence the key rate $K_{\text{coh}} \sim O(\eta_\delta^2)$. In contrast, for the single-photon BB84, $S \sim O(\eta_\delta)$, so that the key rate $K \sim O(\eta_\delta)$. Hence for a given key rate, the use of weak coherent results in a reduced secure distance by a factor of two.

To account for the possible PNS or IRUD attack on QKD schemes using coherent states instead of single photons, protocols that explicitly take into account the multiphoton events were devised. In this section, several of such protocols are briefly described.

3.3.2 *The SARG04 Protocol*

The SARG04, named after its inventors, was tailored to be more robust against PNS attacks [17, 19]. The protocol was based on the observation that BB84 is insecure against PNS attacks whenever Eve can keep one photon because what she needs to do is to discriminate between two eigenstates of a known Hermitian operator after the sifting phase. Therefore, SARG04 works by encoding the classical bit into pairs of nonorthogonal states that cannot be discriminated deterministically. This shares some similarity to the B92 protocol.

In the actual implementation, SARG04 modifies the BB84 protocol by encoding each bit in the basis instead of in the state. In Step 5 of BB84, instead of disclosing the bases Alice used in her measurements, Bob discloses his measurement results x_B. For a particular measurement of Bob, if Alice's preparation basis is the same as his measurement basis, he gets the correct bit value. If Alice's basis is different from his basis, then there is a half chance that the measurement gives an incorrect bit value. Therefore, in Step 6, Alice checks which of Bob's measured bit results x_B are not the same as her bit data x_A and tell Bob to retain only those events. Those events correspond to anti-correlation between the bit sequence encoding Alice's preparation basis y_A and that encoding Bob's measurement basis y_B. Bob just needs to negate the bit sequence y_B. The rest of the protocol are identical to that of BB84. An example is shown in Table 3.3.

Table 3.3 Example of the working of the SARG04 protocol

Step	Alice	Transmission									Bob
1	Bit sequence x_A	1	1	0	0	1	0	1	0	1	
	Bit sequence y_A for basis	1	0	0	0	0	1	1	1	1	
2	Preparation basis	×	+	+	+	+	×	×	×	×	
	Qubit sequence	↘	↕	↔	↔	↕	↗	↘	↗	↘	
4		1	1	0	1	0	0	1	1	0	Bit sequence y_B for basis
		×	×	+	×	+	+	×	×	+	Measurement basis
		1	1	0	1	1	1	1	0	1	Measured bit values x_B
6	Result matching ($x_A = x_B$?)	✓	✓	✓	✗	✓	✗	✓	✓	✓	Result matching ($x_A = x_B$?)
7	Sifted key y'_A	–	–	–	0	–	1	–	–	–	Sifted key $y'_B = y'_A$

The SARG04 protocols in a sense combines the ideas of BB84 and B92. It improves the security against PNS attacks by taking advantage of the fact that Eve can only extract full information from the pulses containing more than 2 photons [17, 19]. The critical attenuation δ_c now is determined by p_3 instead of p_2 as in Eq. (3.17). It turns out to be more resilient to PNS attacks than BB84 by extending the secure distance from ~50 km up to ~100 km in the case of zero errors. In addition, decoy states can be introduced to gain additional knowledge about Eve's attack as described next.

3.3.3 The Decoy-State Method

As discussed earlier, QKD using coherent states can still be secure against the PNS attacks if the attenuation of the quantum channel is low enough and the expected detection at Bob's side is well characterized [see Eqs. (3.17) and (3.18)]. The use of decoy states was thus proposed to characterize the channel so as to gain additional knowledge about Eve's attack and hence increase the distance of the channel that is deemed secure [18, 20, 21].

The secure key rate in Eq. (3.20) can be equivalently written as

$$K_{\text{coh}} = \max\{S_1[1 - h(e_1)] - S_\mu h(Q_\mu), 0\}, \tag{3.26}$$

where S_1 is the gain of Bob's detection events originated from single-photon signals emitted by Alice and e_1 is the QBER of the single photon events. While the overall gain S_μ and overall QBER Q_μ are directly measurable in the protocol, their single-photon counterparts S_1 and e_1 are not. As shown in Eq. (3.21), the worst-case scenario for the estimation of S_1 leads to a very small mean intensity μ that is related to the loss of the channel. The decoy-state method on the other aims to give an upper bound of Δ by measuring the properties of the channel, so as to attain a better value of S_1.

The decoy-state method proceeds by sending decoy pulses with mean intensity μ' with $\mu' \neq \mu$. A main argument of the decoy-state method is that Eve cannot distinguish a decoy state from a signal state, and the only information available to her is the number of photons in the signal. As a consequence, the counting rates or yields in Eq. (3.23) are the same for both signal and decoy signals, i.e., $Y_n(\text{signal}) = Y_n(\text{decoy}) = Y_n$. In the following, we establish the upper bound of Δ following the simple approach of Wang [21].

Suppose Alice can send signals out of three classes of coherent states with mean intensity 0, μ, and μ'. The first and third classes are the decoys (mean intensity 0 is equivalent to empty pulses), whereas the second class is used for the signals. Assume also $\mu' > \mu$ for convenience. With the phase of the coherent states randomized, the quantum state corresponding to the first class is

$$\rho_0 = |0\rangle\langle 0|, \tag{3.27}$$

and that of the μ class is

$$\rho_\mu = e^{-\mu}|0\rangle\langle 0| + \mu e^{-\mu}|1\rangle\langle 1| + c\rho_c, \tag{3.28}$$

where $c = \sum_{n\geq 2} p_n(\mu) = 1 - (1+\mu)e^{-\mu} > 0$ is the multiphoton probability for the class, and

$$\rho_c = \frac{1}{c}\sum_{n=2}^{\infty} p_n(\mu)|n\rangle\langle n|, \tag{3.29}$$

with p_n given in Eq. (3.14), is the multi-photon contribution in ρ_μ. On the other hand, the quantum state corresponding to the μ' class can be written as

$$\rho_{\mu'} = e^{-\mu'}|0\rangle\langle 0| + \mu' e^{-\mu'}|1\rangle\langle 1| + c\frac{\mu'^2 e^{-\mu'}}{\mu^2 e^{-\mu}}\rho_c + d\rho_d, \tag{3.30}$$

where

$$d = \sum_{n\geq 2} p_n(\mu') - \frac{p_2(\mu')}{p_2(\mu)}\sum_{n\geq 2} p_n(\mu) = 1 - (1+\mu')e^{-\mu'} - c\frac{\mu'^2 e^{-\mu'}}{\mu^2 e^{-\mu}}, \tag{3.31}$$

and

$$\rho_d = \frac{1}{d}\left[\sum_{n=2}^{\infty} p_n(\mu')|n\rangle\langle n| - c\frac{\mu'^2 e^{-\mu'}}{\mu^2 e^{-\mu}}\rho_c\right]. \tag{3.32}$$

The way Eq. (3.30) is written so that the μ' class state $\rho_{\mu'}$ is in a form comparable to the μ class state ρ_μ except for the modified probabilities and the extra term ρ_d. It can be shown that $d > 0$:

$$d = p_2(\mu')\sum_{n\geq 2}\left[\frac{p_n(\mu')}{p_2(\mu')} - \frac{p_n(\mu)}{p_2(\mu)}\right] = p_2(\mu')\sum_{n\geq 2}\frac{2}{n!}\left[\mu'^{n-2} - \mu^{n-2}\right] > 0, \tag{3.33}$$

by the assumption of $\mu' > \mu$. Also ρ_d is a proper density matrix:

$$\rho_d = \frac{p_2(\mu')}{d}\sum_{n=2}^{\infty}\left[\frac{p_n(\mu')}{p_2(\mu')} - \frac{p_n(\mu)}{p_2(\mu)}\right]|n\rangle\langle n| > 0, \tag{3.34}$$

and $\mathrm{Tr}\rho_d = 1$. With the convex forms of the density matrices in Eqs. (3.27), (3.28), and (3.30), the action that Alice chooses to send pulses from the three classes randomly is equivalent to have Alice sometimes sending nothing ($|0\rangle\langle 0|$), sometimes one photon ($|1\rangle\langle 1|$), sometimes the multiphoton state ρ_c, and sometimes the multiphoton state ρ_d. Then the counting rates for the three classes are given by

$$S_0 = Y_0(0), \tag{3.35a}$$

$$S_\mu = e^{-\mu} Y_0(\mu) + \mu e^{-\mu} Y_1(\mu) + c Y_c(\mu), \tag{3.35b}$$

$$S_{\mu'} = e^{-\mu'} Y_0(\mu') + \mu' e^{-\mu'} Y_1(\mu') + c \frac{\mu'^2 e^{-\mu'}}{\mu^2 e^{-\mu}} Y_c(\mu') + d Y_d(\mu'), \tag{3.35c}$$

where $Y_\sigma(\mu)$ is the counting rate of the state σ with mean intensity μ. As mentioned previously, Eve cannot distinguish a decoy state from a signal state. Therefore,

$$Y_0(0) = Y_0(\mu) = Y_0(\mu') = S_0, \tag{3.36a}$$

$$Y_1(\mu) = Y_1(\mu') \equiv Y_1, \tag{3.36b}$$

$$Y_c(\mu) = Y_c(\mu') \equiv Y_2. \tag{3.36c}$$

Therefore, Eq. (3.35) can be written as

$$S_\mu = e^{-\mu} S_0 + \mu e^{-\mu} Y_1 + c Y_c, \tag{3.37a}$$

$$S_{\mu'} = e^{-\mu'} S_0 + \mu' e^{-\mu'} Y_1 + c \frac{\mu'^2 e^{-\mu'}}{\mu^2 e^{-\mu}} Y_c + d Y_d. \tag{3.37b}$$

It is remarked that S_0, S_μ, and $S_{\mu'}$ are measurable quantities whereas Y_1, Y_c, and Y_d are unknown.

The tag ratio in Eq. (3.21) is given by the multiphoton events of the signal states, i.e.,

$$\Delta = \frac{c Y_c}{S_\mu}. \tag{3.38}$$

Using Eq. (3.37),

$$c Y_c = \frac{\mu^2 e^{-\mu}}{\mu'^2 e^{-\mu'}} \left(S_{\mu'} - e^{-\mu'} S_0 - \mu' e^{-\mu'} Y_1 - d Y_d \right). \tag{3.39}$$

Therefore,

$$\begin{aligned} \Delta &= \frac{\mu^2 e^{-\mu}}{\mu'^2 e^{-\mu'}} \frac{\left(S_{\mu'} - e^{-\mu'} S_0 - \mu' e^{-\mu'} Y_1 - d Y_d \right)}{S_\mu} \\ &\le \frac{\mu^2 e^{-\mu}}{\mu'^2 e^{-\mu'}} \frac{\left(S_{\mu'} - e^{-\mu'} S_0 \right)}{S_\mu} \le \frac{\mu^2 e^{-\mu}}{\mu'^2 e^{-\mu'}} \frac{S_{\mu'}}{S_\mu}. \end{aligned} \tag{3.40}$$

Therefore, the tag ratio can be upper bounded by the measurable quantities. In normal operations of the protocol that Eve does not disturb the communication,

$$\frac{S_{\mu'}}{S_\mu} = \frac{1 - e^{-\eta_{\mathrm{det}}\eta_\delta \mu'}}{1 - e^{-\eta_{\mathrm{det}}\eta_\delta \mu}} \sim \frac{\mu'}{\mu}. \tag{3.41}$$

As a result,

$$\Delta \leq \frac{\mu e^{-\mu}}{\mu' e^{-\mu'}}. \tag{3.42}$$

Since $\Delta < 1$, the condition required is $\mu' e^{-\mu'} > \mu e^{-\mu}$, together with the assumption $\mu' > \mu$. The main consequence of Eqs. (3.41) and (3.42) is that the choice of μ and the tag ratio are essentially not sensitive to the channel loss, in contrast to the worst case scenario considered when the decoy states are not used. Thus the secure key rate on the channel loss can be recovered to the linear dependence, i.e., $O(\eta_\delta)$.

Finally, a tighter upper bound of Δ can be found by solving for Y_1 and Y_c explicitly in Eq. (3.37) to get [21]

$$\Delta \leq \frac{\mu}{\mu' - \mu}\left(\frac{\mu e^{-\mu} S_{\mu'}}{\mu' e^{-\mu'} S_\mu} - 1\right) + \frac{\mu e^{-\mu} S_0}{\mu' S_\mu}. \tag{3.43}$$

The bound can also be evaluated by using only one decoy state [22].

3.3.4 The COW Protocol

The coherent one way (COW) protocol [23, 24] exploits the phase coherence of consecutive pulses to monitor the properties of the quantum channel when weak coherent pulses are used. In the COW protocol, Alice produces pulses of mean photon-number μ that are separated with a well-defined fixed time period τ. An intensity modulator is used to create either an empty pulse (stop) or a nonempty pulse (go). The kth logical bit is encoded in a two-pulse sequence of the empty and nonempty pulses using the prescription:

$$|0_A\rangle = \left|\sqrt{\mu}\right\rangle_{2k-1} |0\rangle_{2k}, \tag{3.44a}$$

$$|1_A\rangle = |0\rangle_{2k-1} \left|\sqrt{\mu}\right\rangle_{2k}, \tag{3.44b}$$

where $\left|\sqrt{\mu}\right\rangle$ is a coherent state with amplitude $\sqrt{\mu}$ and $|0\rangle$ is the vacuum state with no photon (see Chap. 2). Note that $|0_A\rangle$ and $|1_A\rangle$ are not orthogonal. Nevertheless, these two states can be unambiguously discriminated optimally by measuring the arrival time of the pulses. To check channel characteristics, Alice also sends a tiny portion of decoy sequence $|\sqrt{\mu}\rangle_{2k-1}|\sqrt{\mu}\rangle_{2k}$ or as $|\sqrt{\mu}\rangle_{2k}|\sqrt{\mu}\rangle_{2k+1}$ across a bit separation when a sequence $|1\rangle_k|0\rangle_{2k+1}$ (logical bit 0) is coded. Since the laser is mode-locked, there is a phase coherence between any two non-empty pulses (see Fig. 3.1).

In Fig. 3.1, at Bob's side, the pulses are split into two paths: (1) transmitted to detector D_B (dataline) and (2) reflected into the interferometer comprising of detectors D_{M1} and D_{M2} (monitoring line). For the dataline, D_B measures the pulse intensities and thus can distinguish among the logical bits and decoys. For the monitoring line, when both pulses at time k and $k+1$ are nonempty, only D_{M1} can fire at time $k+1$. If Eve breaks the coherence by reading the channel, she can be detected by the firing at D_{M2} for those events. Consider the three right-most pulses in Fig. 3.1. The detection of Eve now reads as: if she attacks coherently across the bit separation (pulses 2 and 3), she breaks the coherence of the decoy sequences (pulses 1 and 2); if she attacks coherently within each bit (pulses 1 and 2), then she breaks the coherence across the separation (pulses 2 and 3). Eve can make a coherent attack on a larger number of pulses, but in this case she breaks the coherence in fewer positions but gets much less information.

The steps of COW is given below:

1. Alice sends a large number of pulses. Most of them correspond to the logical bits 0 and 1. A small portion of them correspond to the decoys.
2. Bob announces that he received the qubits through the public classical channel. Then Alice reveals the time slot $\{k_d\}$ corresponding to a decoy sequence. Bob removes all the detections at times $2k_d - 1$ and $2k_d$. He also checks whether detector D_{M2} has ever fired at time $2k_d$. Then Alice and Bob can estimate the break of coherence of the decoy pulses by calculating the visibility

$$V = \frac{p(D_{M1}) - p(D_{M2})}{p(D_{M1}) + p(D_{M2})}, \tag{3.45}$$

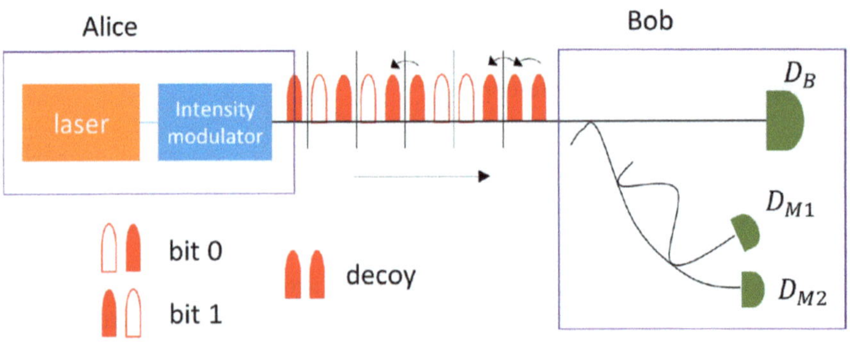

Fig. 3.1 Schematic of the COW protocol. Arrows over the pulses denote coherence

where $p(D_{Mj})$ is the probability of detection in D_{Mj} at the time only D_{M1} should have fired. The visibility should be unity if Eve did not attack.

3. For the bits that do not correspond to the decoy sequences, Bob reveals the times $2k+1$ that he observed detection in D_{M2}. Alice checks if any of those correspond to the logical bit 0 event $|1\rangle_k|0\rangle_{k+1}$. Then again Alice and Bob can estimate the break of coherence across the bit separation.
4. Based on Steps 2 and 3, Alice and Bob can computes the information that could have been leaked to Eve.
5. Alice and Bob perform error correction and privacy amplification to come up with the final secret key.

The visibility can be used to calculate the information leaked to Eve based on the visibilities and the bit error rate. For details, see [23, 24].

3.4 Entangled Photon-Based QKD Protocol

The QKD protocols described in the previous sections are usually called prepare-and-measure (P&M) protocols which describe the basic steps of the protocols. In a somewhat different direction, the E91 protocol was proposed independently in 1991 by Ekert [25] that exploited the exotic notion of quantum entanglement, which was central to the interpretation and the studies of the completeness of the theory of quantum mechanics in its early development.

In Chap. 2, we mention that a quantum system is completely described by its quantum state. What is so peculiar about the quantum state is that one cannot observe the quantum state directly—it has to be done only indirectly through some measurement procedure in accordance to the postulates of quantum mechanics. The measurement nevertheless changes the quantum state if the state is not an eigenstate of the measurement. This notion somewhat is also manifested in the no-cloning theorem. This can be interpreted as: the outcomes of the measurements did not exist before the measurements. Worse still, when the measurement involves a pair of entangled particles which are strongly correlated in certain physical attributes, measurement of the attribute of one particle can modify the attribute of the other particle, even though they may be separated from each other by a large distance. This is the scenario exhibited by the famous Einstein-Podolsky-Rosen (EPR) paradox [26]. As a consequence, before any measurement, there is no information carried by the entangled particles (property 1). More importantly, any disturbance to the entangled particles degrades or even breaks the correlation between the two particles (property 2). Entanglement-based QKD schemes make use of these two properties to establish secure keys with the schemes capable of detecting the presence of the eavesdropper.

In this section, the two properties are first made more definite through the Bell state introduced in Chap. 2 and the Bell's inequality. The E91 protocol is then described.

3.4.1 *Quantum Entanglement and Bell's Inequality*

Consider the Bell state

$$|\Psi^-\rangle^{AB} = \frac{|01\rangle^{AB} - |10\rangle^{AB}}{\sqrt{2}}, \tag{3.46}$$

which is an entangled state of two qubits. Also consider the POVM $\{E_0 = |0\rangle\langle 0|, E_1 = |1\rangle\langle 1|\}$, i.e., the eigenkets of Z operator. It can be seen that the only measurement outcomes with nonzero probabilities are

$$P(A = 0, B = 1|\Psi^-) = {}^{AB}\langle\Psi^-|E_0^A E_1^B|\Psi^-\rangle^{AB} = \frac{1}{2}, \tag{3.47a}$$

$$P(A = 1, B = 0|\Psi^-) = {}^{AB}\langle\Psi^-|E_1^A E_0^B|\Psi^-\rangle^{AB} = \frac{1}{2}, \tag{3.48a}$$

so that Alice's and Bob's events are anti-correlated in the basis of Z. On the other hand, it can be proven that

$$|\Psi^-\rangle^{AB} = \frac{|+-\rangle^{AB} - |-+\rangle^{AB}}{\sqrt{2}}, \tag{3.48b}$$

by changing to the Hadamard basis. Therefore, their events are also anti-correlated in the basis of X. In fact, it can be shown that

$$|\Psi^-\rangle^{AB} = e^{i\theta}\frac{|xy\rangle^{AB} - |yx\rangle^{AB}}{\sqrt{2}}, \tag{3.49}$$

with

$$|x\rangle = \alpha|0\rangle + \beta|1\rangle, \tag{3.50a}$$

$$|y\rangle = \gamma|0\rangle + \delta|1\rangle, \tag{3.50b}$$

where

$$e^{i\theta} = \alpha\delta - \beta\gamma = \begin{vmatrix} \alpha & \beta \\ \gamma & \delta \end{vmatrix}. \tag{3.50c}$$

The interpretation is that, before Alice and Bob make measurement, the Bell state does not have definite information. Moreover, when the measurement basis is fixed, Alice's measurement outcome fixes that of Bob, even though they can be separated by a large distance. It should nevertheless be noted that there is no violation of special relativity or causality, because Eq. (3.47) involves joint measurements that require verifying the joint outcomes by bringing the two results together, which is under the limit of the speed of light.

Historically, the Bell state (3.46) is a discrete variable version of the EPR state involving continuous position and momentum correlations in the EPR paradox [26]. In 1964, John Bell found that the EPR paradox can be experimentally verified using the Bell state through the Bell's inequality [27, 28]. Its modified version, called the CHSH inequality named after the initials of its four discoverers, is explained in the following [29].

Consider the experiment in which two particles are prepared with some physical attributes that may be correlated. The two particles are each received by Alice and Bob, who are separated from each other. On Alice side, suppose she can perform two different sets of measurements, labelled by P_Q and P_R respectively, on the particle she receives, and she does not know in advance which set of measurement she will perform. Suppose further each of the measurement set gives rise to two possible outcomes, labelled by the values $+1$ and -1. Thus, we write $Q = \pm 1$ and $R = \pm 1$. Likewise, Bob can perform similar measurements with the sets P_S and P_T, which also give the outcomes $S = \pm 1$ and $T = \pm 1$. Now using the values of the outcomes, we can find that

$$QS + RS + RT - QT = (Q + R)S + (R - Q)T. \tag{3.51}$$

Note that given any combination for Q and R with $Q, R = \pm 1$, either $Q + R$ or $R - Q$ vanishes. For the case that does not vanish, it must be either $+2$ or -2. As a consequence, $QS + RS + RT - QT = \pm 2$, for $S, T = \pm 1$, no matter what the values of S and T are. When this experiment is repeated many times, the expectation value of Eq. (3.51) is given by, using classical probability theory,

$$\begin{aligned} \mathbf{E}(QS + RS + RT - QT) &= \sum_{q,r,s,t=\pm 1} p(q,r,s,t)(qs + rs + rt - qt) \\ &\leq 2 \sum_{q,r,s,t=\pm 1} p(q,r,s,t) = 2. \end{aligned} \tag{3.52}$$

On the other hand,

$$\begin{aligned}\mathbf{E}(QS+RS+RT-QT) &= \sum_{q,r,s,t=\pm1} p(q,r,s,t)(qs+rs+rt-qt)\\ &= \sum_{q,r,s,t=\pm1} p(q,r,s,t)(qs) + \sum_{q,r,s,t=\pm1} p(q,r,s,t)(rs)\\ &\quad + \sum_{q,r,s,t=\pm1} p(q,r,s,t)(rt) - \sum_{q,r,s,t=\pm1} p(q,r,s,t)(qt)\\ &= \mathbf{E}(QS)+\mathbf{E}(RS)+\mathbf{E}(RT)-\mathbf{E}(QT).\end{aligned} \tag{3.53}$$

Thus,

$$-2 \le \mathbf{E}(QS)+\mathbf{E}(RS)+\mathbf{E}(RT)-\mathbf{E}(QT) \le 2. \tag{3.54}$$

It is remarked that Eq. (3.54) is obtained based on the general probability $p(q,r,s,t)$, which can describe any classical correlation among the random variables Q,R,S,T.

Now suppose the two particles are two qubits described by the Bell state $|\Psi^-\rangle$ in Eq. (3.46). The measurements performed by Alice and Bob correspond to the following observables:

$$Q = Z^A, \quad R = X^A, \quad S = -\frac{Z^B + X^B}{\sqrt{2}}, T = \frac{Z^B - X^B}{\sqrt{2}}. \tag{3.55}$$

Equation (3.55) can be written as

$$Q = \vec{q}\cdot\vec{\sigma}, \quad R = \vec{r}\cdot\vec{\sigma}, \quad S = \vec{s}\cdot\vec{\sigma}, \quad T = \vec{t}\cdot\vec{\sigma}. \tag{3.56}$$

where $\vec{\sigma} = X\vec{x} + Y\vec{y} + Z\vec{z}$, with X,Y,Z being the Pauli matrices defined in Eqs. (2.33), (2.36), and (2.37), and

$$\vec{q} = \vec{z}, \tag{3.57a}$$

$$\vec{r} = \vec{x}, \tag{3.57b}$$

$$\vec{s} = -\frac{\vec{x}+\vec{z}}{\sqrt{2}}, \tag{3.57c}$$

$$\vec{t} = \frac{-\vec{x}+\vec{z}}{\sqrt{2}}. \tag{3.57d}$$

Note that $\vec{q},\vec{r},\vec{s},\vec{t}$ are unit vectors and are pictorially shown in Fig. 3.2. In addition, an operator of the form $\vec{a}\cdot\vec{\sigma}$ with $\vec{a}$ a unit vector can be shown to have eignevalues ±1. Explicitly, let $\vec{a} = \left(a_x, a_y, a_z\right)$. Then

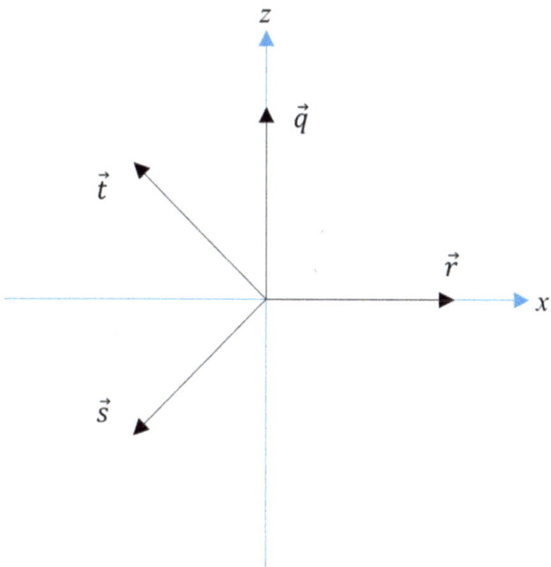

Fig. 3.2 Illustration of the vectors corresponding to the four quantum measurements in the violation of the CHSH inequality. The sets of vectors $\{\vec{q}, \vec{r}\}$ and $\{\vec{s}, \vec{t}\}$ can be viewed as two sets of orthogonal vectors that are rotated by an angle of $\pi/4$. It is remarked that the vector $\vec{s}$ is equivalent to $-\vec{s}$, only that Eqs. (3.54) and (3.61) need to be changed to $-(\langle QS\rangle - \langle RT\rangle + \langle RS\rangle + \langle QT\rangle)$

$$A = \vec{a}\cdot\vec{\sigma} = \begin{bmatrix} a_z & a_x - ia_y \\ a_x + ia_y & -a_z \end{bmatrix}. \tag{3.58}$$

Therefore, the eigenvalue λ can be obtained by solving

$$0 = |\vec{a}\cdot\vec{\sigma} - \lambda I| = \begin{vmatrix} a_z - \lambda & a_x - ia_y \\ a_x + ia_y & -a_z - \lambda \end{vmatrix} = \lambda^2 - \left(a_x^2 + a_y^2 + a_z^2\right) = \lambda^2 - 1, \tag{3.59}$$

that is, $\lambda = \pm 1$.

The quantum expectation values using the qubits and measurements can be calculated using the relations

$$\langle\Psi^-|X^AX^B|\Psi^-\rangle = \left(\frac{\langle 01| - \langle 10|}{\sqrt{2}}\right)X^AX^B\left(\frac{|01\rangle - |10\rangle}{\sqrt{2}}\right) = -1, \tag{3.60a}$$

$$\langle\Psi^-|Z^AZ^B|\Psi^-\rangle = \left(\frac{\langle 01| - \langle 10|}{\sqrt{2}}\right)Z^AZ^B\left(\frac{|01\rangle - |10\rangle}{\sqrt{2}}\right) = -1, \tag{3.60b}$$

$$\langle\Psi^-|X^AZ^B|\Psi^-\rangle = \left(\frac{\langle 01| - \langle 10|}{\sqrt{2}}\right)X^AZ^B\left(\frac{|01\rangle - |10\rangle}{\sqrt{2}}\right) = 0, \tag{3.60c}$$

$$\langle\Psi^-|Z^AX^B|\Psi^-\rangle = \left(\frac{\langle 01| - \langle 10|}{\sqrt{2}}\right)Z^AX^B\left(\frac{|01\rangle - |10\rangle}{\sqrt{2}}\right) = 0. \tag{3.60d}$$

Therefore, $\langle\Psi^-|QS\langle\Psi^-| = -\vec{q}\cdot\vec{s} = 1/\sqrt{2}$. Likewise, we obtain

$$\langle QS\rangle + \langle RT\rangle + \langle RS\rangle - \langle QT\rangle = -\vec{q}\cdot\vec{s} - \vec{r}\cdot\vec{t} - \vec{r}\cdot\vec{s} + \vec{q}\cdot\vec{t} = 2\sqrt{2}, \tag{3.61}$$

which violates the CHSH inequality (3.54). Such violation stems from the fact that the quantum correlation contained in the Bell state is such that it cannot be described by a classical probability through $p(q,r,s,t)$ and the sets of measurements by Alice and Bob are nonorthogonal (see Fig. 3.2). In experiment, it turns out Eq. (3.61) is correct.

It should be noted that the Bell's inequality makes two basic assumptions: (1) the random variables Q,R,S,T have definite values that are independent of the actual measurements, and (2) Alice's measurements cannot influence Bob's measurements. The former is known as the assumption of realism, while the latter is the assumption of locality. For details, see [2].

3.4.2 *The E91 Protocol*

For quantum cryptography, the main application of the Bell's inequality is that the action of Eve on Bob's qubit breaks the quantum correlation of Alice's and Bob's qubits. Therefore, by checking whether the Bell's inequality is violated or not, Alice and Bob can detect the presence of Eve. The following illustrates the steps of the E91 protocol [24].

1. N Bell states of the form in Eq. (3.46) are prepared. For each of the Bell state, one qubit is given to Alice and the other qubit is sent to Bob.
2. Upon receiving the qubits, Alice makes measurements by randomly choosing from the set

$$A_1 = Q^A, \quad A_2 = R^A, \quad A_3 = S^A, \tag{3.62}$$

and Bob makes measurements by randomly choosing from the set

$$B_1 = S^B, \quad B_2 = T^B, \quad B_3 = Q^B, \tag{3.63}$$

in which Q,R,S,T are defined in the same way as Eq. (3.55). The superscripts A and B are added to denote the measurement performed on Alice's or Bob's qubit. Any of these measurements give a value $+1$ or -1.

3. After their measurements, Alice and Bob announce in the public channel the measurements (not the measurement results) they have chosen. Then they divide the measurements into two groups: (1) they used the same measurements (the pairs $\{A_1, B_3\}$ and $\{A_3, B_1\}$) and (2) they used different measurements (the remaining pairs). They discard all measurements in which either or both of them did not register any detection.
4. Alice and Bob reveal in the public channel the measurement results corresponding to the second groups. In particular, using the pairs $\{A_1, B_1\}, \{A_1, B_2\}, \{A_2, B_1\}, \{A_2, B_2\}$, they can calculate the expectation values in Eq. (3.61). If the channel has not be disturbed, they should be able to obtain the value in Eq. (3.61). This ensures that the measurement results obtained by Alice and Bob are anti-correlated within the second group of measurements.
5. Suppose in her measurement results in the second group, Alice uses $+1$ to represent bit value 0 and -1 to represent bit value 1. Then Bob uses in his measurement results -1 to represent bit value 0 and $+1$ to represent bit value 1. This forms the raw key shared in common by Alice and Bob.
6. The remaining steps are the same as Steps 7–9 of BB84.

In the protocol, Step 4 is used to check the security. The eavesdropper Eve cannot obtain any information when the qubits are sent to either Alice or Bob because there is no information encoded in the qubits [cf. Equation (3.49)]. The information comes into existence only after Alice and Bob make the measurements and communicate in the public channel. If Eve interferes the qubits, the violation against the CHSH inequality will decrease. It can be shown that the amount of information obtained by Eve goes inversely to the degree of violation of the CHSH inequality. In the event that Eve obtains full information, Eq. (3.54) will no longer be violated.

3.5 Challenges of Current Approaches of QKD

Most of the current QKD schemes (BB84, SARG04, COW, etc.) are derived from the BB84 protocol. The low mean photon number per pulse required by BB84 (decoy state or not) and the limited efficiency and performance of single-photon detectors [8] limit the secure key rate currently to around 1 Mbps [30–32], which is orders of magnitude smaller than the data rate of modern fiber optical communication applications at >100 Gbps. BB84-based QKD systems therefore currently mainly focus on generating finite keys for classical cryptographic methods (such as AES256) [33, 34], which though powerful is still prone to quantum computer and cryptanalytic attacks with increasing computational power and undermines the promise of unconditional security afforded by QKD for one-time pad (OTP).

Continuous-variable (CV) QKD systems have been proposed and experimentally demonstrated [35], which use only homodyne or heterodyne measurement of light-field quadratures instead of single-photon detection. CV-QKD nevertheless requires coherent detection with a local oscillator (LO) that potentially leaves loopholes for manipulation by the adversary [36]. The classical post-processing of CV-QKD is a complex operation (performing error-correction at low SNRs) and the large amount of classical data needed for the sifting procedure and reconciliation imposes a limit to the key rates of CV-QKD similar to that of BB84 [35].

Finally, even though QKD protocols can be proven to be theoretically secure, it has been demonstrated that imperfect light sources and detection devices could lead to issues that can compromise the security of QKD systems. For a summary of the various quantum hacking attacks against commercial and research QKD setups, see [32] and the references therein.

3.6 Summary

Quantum key distribution is an active field of research that is still undergoing vigorous development. This chapter has discussed the concept behind QKD and mentioned several most common prepare-and-measure (P&M) and entanglement-based QKD protocols. The decoy version of the discussed P&M QKD protocols have in fact been implemented in commercial QKD systems [33, 34].

It is remarked that the prepare-and-measure and entanglement-based approaches are intimately related to each other [37]. Any prepare-and-measure scheme can be translated into an entanglement-based scheme [4]. In fact, the rigorous security proof of QKD involving the classical post-processing of error correction and privacy amplification relies on the clever use of entanglement. The proof has been omitted in this chapter because it involves details about entanglement distillation and quantum error correction [5, 6].

To cope with the issue of quantum hacking, recent developments in QKD have focused on the so-called device-independent QKD that takes into account practical issues intrinsically related to the theory [32, 38, 39]. Interested readers are referred to the relevant references.

References

1. Katz, J., & Lindell, Y. (2014). *Introduction to modern cryptography* (2nd ed.). Boca Raton: CRC Press.
2. Nielsen, M., & Chuang, I. (2011). *Quantum computation and quantum information: 10th anniversary edition* (10th Anniversary Edition). Cambridge: Cambridge University Press.
3. Bennett, C. H., & Brassard, G. (1984). Quantum cryptography: Public key distribution and coin tossing. In *Proceedings of IEEE International Conference on Computers, Systems and Signal Processing, New York*, Vol. 175, 8pp.
4. Gisin, N., Ribordy, G., Tittel, W., & Zbinden, H. (2002). Quantum cryptography. *Review of Modern Physics, 74,* 145.

5. Shor, P. W., & Preskill, J. (2000). Simple proof of security of the BB84 quantum key distribution protocol. *Physical Review Letters, 85,* 441.
6. Gottesman, D., Lo, H. K., Lütkenhaus, N., & Preskill, J. (2004). Security of quantum key distribution with imperfect devices. *Quantum Information and Computation, 4,* 325.
7. Bennett, C. H. (1992). Quantum cryptography using any two nonorthogonal states. *Physical Review Letters, 68,* 3121.
8. Eisaman, M. D., Fan, J., Migdall, A., & Polyakov, S. V. (2011). Invited review article: Single-photon sources and detectors. *Review of Scientific Instruments, 82,* 071101.
9. Buckley, S., Rivoire, K., & Vučković, J. (2012). Engineered quantum dot single-photon sources. *Reports on Progress in Physics, 75,* 126503.
10. Takemoto1, K., Nambu, Y., Miyazawa, T., Sakuma, Y., Yamamoto1, T., Yorozu, S., & Arakawa, Y. (2015). Quantum key distribution over 120 km using ultrahigh purity single-photon source and superconducting single-photon detectors. *Scientific Reports, 5,* 14393.
11. Brassard, G., Lütkenhaus, N., Mor, T., & Sanders, B. C. (2000). Limitations on practical quantum cryptography. *Physical Review Letters, 85,* 1330.
12. Grangier, P., Levenson, J. A., & Poizat, J. P. (1998). Quantum non-demolition measurements in optics. *Nature, 396,* 537.
13. Johnson, B. R., Reed, M. D., Houck, A. A., Schuster, D. I., Bishop, Lev S., Ginossar, E., et al. (2010). Quantum non-demolition detection of single microwave photons in a circuit. *Nature Physics, 6,* 663.
14. Lvovsky, A. I., Sanders, B. C., & Tittel, W. (2009). Optical quantum memory. *Nature Photonics, 3,* 706.
15. Vernaz-Gris, P., Huang, K., Cao, M., Sheremet, A. S., & Laurat, J. (2018). Highly-efficient quantum memory for polarization qubits in a spatially-multiplexed cold atomic ensemble. *Nature Communications, 9,* 363.
16. Dušek, M., Jahma, M., & Lütkenhaus, N. (2000). Unambiguous state discrimination in quantum cryptography with weak coherent states. *Physical Review A, 62,* 022306.
17. Scarani, V., Acín, A., Ribordy, G., & Gisin, N. (2004). Quantum cryptography protocols robust against photon number splitting attacks for weak laser pulse implementations. *Physical Review Letters, 92,* 057901.
18. Lo, H. K., Ma, X., & Chen, K. (2005). Decoy state quantum key distribution. *Physical Review Letters, 94,* 230504.
19. Acín, A., Gisin, N., & Scarani, V. (2004). Coherent-pulse implementations of quantum cryptography protocols resistant to photon-number-splitting attacks. *Physical Review A, 69,* 012309.
20. Hwang, W. Y. (2003). Quantum key distribution with high loss: Toward global secure communication. *Physical Review Letters, 91,* 057901.
21. Wang, X. B. (2005). Beating the photon-number-splitting attack in practical quantum cryptography. *Physical Review Letters, 94,* 230503.
22. Ma, X., Qi, B., Zhao, Y., & Lo, H. K. (2005). Practical decoy state for quantum key distribution. *Physical Review A, 72,* 012326.
23. Gisin, N., Ribordy, G., Zbinden, H., Stucki, D., Brunner, N., & Scarani, V. (2004). Towards practical and fast quantum cryptography. arXiv:quant-ph/0411022.
24. Stucki, D., Brunner, N., Gisin, N., Scarani, V., & Zbinden, H. (2005). Fast and simple one-way quantum key distribution. *Applied Physics Letters, 87,* 194108.
25. Ekert, A. K. (1991). Quantum cryptography based on Bell's theorem. *Physical Review Letters, 67,* 661.
26. Einstein, A., Podolsky, B., & Rosen, N. (1935). Can quantum-mechanical description of physical reality be considered complete? *Physical Review, 47,* 777.
27. Bell, J. (1964). On the Einstein Podolsky Rosen Paradox. *Physics, 1,* 195.
28. Bell, J. (1987). *Speakable and unspeakable in quantum mechanics*. Cambridge: Cambridge University Press.

29. Clauser, J., Horne, M., Shimony, A., & Holt, R. (1969). Proposed experiment to test local hidden-variable theories. *Physical Review Letters, 23,* 880.
30. Dixon, A. R., Yuan, Z. L., Dynes, J. F., Sharpe, A. W., & Shields, A. J. (2008). Gigahertz decoy quantum key distribution with 1 Mbit/s secure key rate. *Optics Express, 16,* 18790.
31. Dixon, A. R., Yuan, Z. L., Dynes, J. F., Sharpe, A. W., & Shields, A. J. (2010). Continuous operation of high bit rate quantum key distribution. *Applied Physics Letters, 96,* 161102.
32. Lo, H. K., Curty, M., & Tamaki, K. (2014). Secure quantum key distribution. *Nature Photonics, 8,* 595.
33. http://www.idquantique.com/quantum-safe-crypto/qkd-overview/.
34. http://qubitekk.com/security/.
35. Jouguet, P., Kunz-Jacques, S., Leverrier, A., Grangier, P., & Diamanti, E. (2013). Experimental demonstration of long-distance continuous-variable quantum key distribution. *Nature Photonics, 7,* 378.
36. Huang, J. Z., Weedbrook, C., Yin, Z. Q., Wang, S., Li, H. W., Chen, W., et al. (2013). Quantum hacking of a continuous-variable quantum-key-distribution system using a wavelength attack. *Physical Review A, 87,* 062329.
37. Bennett, C. H., Brassard, G., & Mermin, N. D. (1992). Quantum cryptography without Bell's theorem. *Physical Review Letters, 68,* 557.
38. Mayers, D., & Yao, A. (1998). Quantum cryptography with imperfect apparatus. In *Proceedings of the 39th Annual Symposium on Foundations of Computer Science, Palo Alto, 1998*, p. 503. Washington, DC: IEEE.
39. Vazirani, U., & Vidick, T. (2014). Fully device-independent quantum key distribution. *Physical Review Letters, 113,* 140501.

Chapter 4
Secure Communication Based on Quantum Noise

Advantage creation through intrusion-level detection used by BB84-based QKD protocols is only one possibility permitted by quantum effects. In the early 2000s, another class of quantum cryptography protocols, called keyed communication in quantum noise (KCQ) based on quantum detection and communication theory, was proposed. A main advantage of the KCQ protocols is that they generally allow the use of hundreds or thousands of photons in a signal pulse in contrast to the nominally single photon per pulse in BB84-based QKD protocols. This chapter introduces the concept of KCQ and describes certain implementations of the KCQ protocol.

4.1 Introduction

In conventional symmetric cryptography, it is well-known that the one-time pad (OTP) is theoretically unbreakable. It is, however, inefficient because the length of the pre-shared key needs to be as long as the data. Quantum key distribution (QKD), which addresses the issue of distributing secure keys for symmetric-key encryption, is the most promising aspect of quantum cryptography. The scheme with QKD + OTP thus can provide the strongest security at the physical layer [1, 2]. Nevertheless, QKD suffers from technological challenges as briefly mentioned in Chap. 3 that hinder its widespread deployment. On the other hand, a new category of multi-photon quantum communication represented by the keyed communication in quantum noise (KCQ) [3] has been proposed and implemented to address the stringent technological requirements of BB84-based QKD.

As described in Chap. 3, BB84-based QKD relies on the no-cloning theorem and the use of single photons, which guarantee that the eavesdropper cannot gain information without making disturbances and hence being detected. More precisely, advantage creation is obtained through intrusion-level detection that quantitatively assures the eavesdropper's observation to be inferior to the users', so that classical

P. K. Verma et al., *Multi-photon Quantum Secure Communication*,
Signals and Communication Technology,
https://doi.org/10.1007/978-981-10-8618-2_4

post-processing (i.e., privacy amplification) can be used to eliminate the eavesdropper's information on the final key.

Different from the usual QKD, KCQ relies on the asymmetry in the optimal quantum measurements between the legitimate user and the adversary. More specifically, it imposes irreducible quantum random noise on the eavesdropper's measurements that is much larger than the noise perceived by the legitimate users, which is a practical realization of the wiretap channel [4]. An obvious advantage of this approach is the ability to use tens, hundreds, or even thousands of photons per pulse versus the nominally single photon in BB84. It can be used directly for data encryption at data rates >10 Gbps with security stronger than classical methods. KCQ can be readily implemented in the existing optical communication infrastructure including the use of optical signal amplification in a wave-length-division-multiplexed fiber-optic network [5]. In addition, it can, in principle, be used in a multiple-point to multiple-point topology, such as a passive optical network (PON), in contrast to the point-to-point setting with QKD which also requires quantum repeaters in order to form a network.

4.2 Keyed Communication in Quantum Noise (KCQ)

The basic idea of KCQ is to utilize a shared secret key between the users to determine the quantum signal set that encodes the information to be communicated. In classical communication, anyone tapping into the communication channel, including the adversary can, in principle, get an identical copy of the data the legitimate user receives. In contrast, KCQ exploits that fact that the structure of a quantum receiver that delivers the optimal performance depends on knowledge of the signal set, i.e., with and without the key. As a result, adversaries who do not know the shared key make measurements that are noisier (due to quantum noise) compared to those who possess the key, and hence the advantage created for the latter. The measured data of the illegitimate user is unavoidably different from that of the legitimate user even though the former may learn the key later. This consequence is significant because it allows fresh key to be generated that is much larger than the secret key used during key generation.

The use of quantum noise in measurement has also been explored to directly perform key exchange without a pre-shared key, though care has to be taken to address the man-in-the-middle (MIM) attack [6, 7]. This will be discussed in more detail in later chapters of this book. It is remarked that unconditionally secure key exchange based on classical noise is also possible [8]. In particular, a protocol utilizing Johnson noise and Kirchhoff's law has been proposed [9], which nevertheless relies on electric signals on copper wires for communication and the scheme seems to have very poor security performance in terms of maximal distance ($\sim$1 km) and data rate ($\sim$1 kHz). It should also be emphasized that a shared secret key is needed in *all* known key generation protocols, classical or quantum. For BB84-type protocols, this key is needed in order to authenticate the sender of the message. Without authentication, the adversary can launch the MIM attack to pretend to be the legitimate receiver to the sender and the legitimate sender to the receiver. The shared secret key in KCQ, on the other hand, plays a more essential role than authentication. In addition, since KCQ relies on bright coherent states without the use of other quantum resources such as entangled photons, as in a comparable high-rate floodlight QKD system [10], it has the potential of being applied to wireless systems in addition to optical systems [11–13].

The essential steps of a KCQ key generation protocol are [14]:

(i) The use of a shared secret key K between Alice and Bob to choose the quantum states to encode the data bit sequences (by Alice) and to determine the measurement basis to detect the quantum states (by Bob) that gives them a better error performance over the intruder Eve who has no knowledge of K when she makes her measurements;
(ii) A method for Alice and Bob to extract a fresh key using the performance advantage described;
(iii) A key verification process possibly using another shared secret key K_v between Alice and Bob.

There can be different implementations of KCQ. A specific scheme is detailed in the following.

4.2.1 KCQ Coherent-State Key Generation with Binary Detection

In this protocol, also called the $\alpha\eta$ protocol (or Y00 protocol by researchers in Japan) for direct encryption [15, 16], information is encoded in qumodes (quantum modes) instead of in qubits [3, 14, 15]. A qumode is a coherent state of an infinite-dimensional space. In the protocol, there are $2M$ possible states. In the single-mode realization, the qumodes are coherent states of the form

$$|\alpha_l\rangle = \left|\alpha e^{i\theta_l}\right\rangle, \quad \theta_l = \tfrac{2\pi l}{2M}, \quad l \in \{1, 2, \ldots, 2M\}, \tag{4.1}$$

where α^2 is the mean photon number in the state. The qumodes can also be represented in a two-mode realization (such as the polarizations):

$$|\alpha_l\rangle = |\alpha \cos\theta_l\rangle_1 |\alpha \sin\theta_l\rangle_2. \tag{4.2}$$

The qumodes may be associated with polarization [15], phase shift [5], or any type of physical mode of photons. An intensity-modulation scheme for the protocol has also been demonstrated [16].

In the binary detection scheme, the $2M$ qumodes are grouped as M pairs $\alpha_l \equiv \{|\alpha_l\rangle, |\alpha_{l+M}\rangle\}$ which are used to encode the bit values 0 and 1 (or 1 and 0). The choice of which pair to use for a particular pulse is determined by $\log M$ bits from a random sequence produced using a pseudorandom-number generator (PRNG). Figure 4.1 depicts the overall scheme of the $\alpha\eta$ cryptosystem.

According to the KCQ model, Alice and Bob share a secret seed key K, with the length $|K| = 100{-}1000$ bits. The key is extended to a running key (M-ary sequence) K_r by a PRNG with length $|K_r| < \left(2^{|K|} - 1\right)/\log M$. Each number in the running key is then mapped to a qumode α_l according to a predefined mapper. For example, the regular mapper (RM) $\mathrm{map}(x) = x$ is given by

$$\begin{pmatrix} k_j \\ \alpha_{\mathrm{map}(k_j)} \\ \Pi_{\mathrm{map}(k_j)} \end{pmatrix} = \begin{pmatrix} 1 & 2 & 3 & 4 & \ldots & M \\ \alpha_1 & \alpha_2 & \alpha_3 & \alpha_4 & \ldots & \alpha_M \\ 0 & 1 & 0 & 1 & \ldots & 0 \end{pmatrix}, \tag{4.3}$$

in which the parity Π is added to the final qumode to mix the 0 and 1 bit value (see Fig. 4.1). Given the plaintext $X = x_1, x_2, \ldots x_n$, the transmitted quantum state is a sequence of qumodes

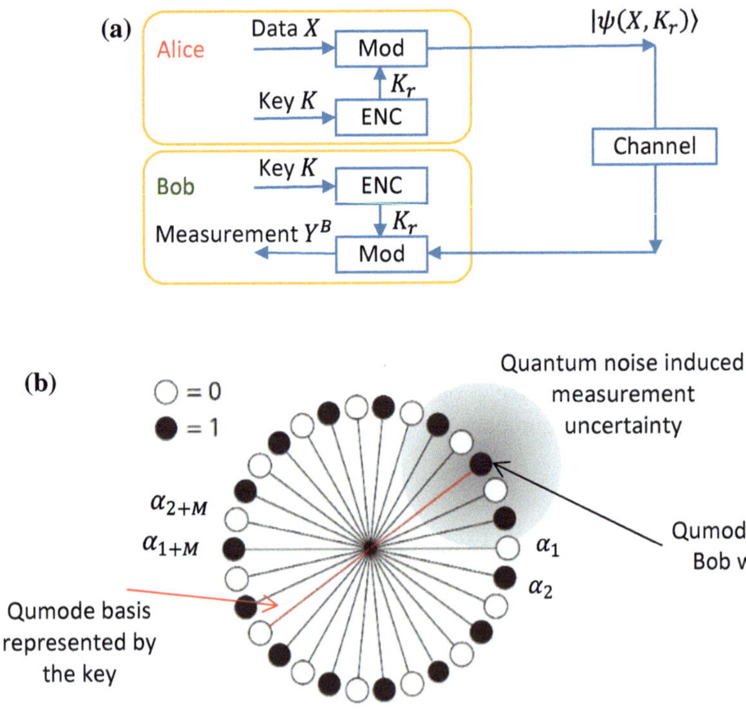

Fig. 4.1 **a** Schematic of the αη cryptosystem. ENC denotes the PRNG with a mapper that drives the modulator (Mod) for the qumodes. **b** Phase-space representation of qumodes (M = 15). A large M is usually used so that quantum noise in Eve's measurement conceals th the actual qumode used. In the figure, the number of qumodes under the masking effect is 5. The two states for Bob to distinguish (the two ends of the qumode basis in red) are well separated even with measurement noise

$$|\psi(X, K_r)\rangle = \left|\alpha_{k_1 + M\left(x_1 + \Pi_{k_1}\right)}\right\rangle \left|\alpha_{k_1 + M\left(x_2 + \Pi_{k_2}\right)}\right\rangle \cdots \left|\alpha_{k_1 + M\left(x_n + \Pi_{k_n}\right)}\right\rangle, \tag{4.4}$$

with the running key $K_r = k_1,\ k_2,\ \ldots k_n,\ k_j \in \{1, \ldots, M\}$. The regular mapper is usually not used as the quantum noise in the detection process affects differently the $\log M$ bits representing the random number for a particular chosen qumode. Such an effect could be exploited by a fast correlation attack [17]. Instead a proper mapper to spread the noise to different bits needs to be employed. For example, one can design an irregular mapping such as [18]

$$2m - \delta \leq d_H(\alpha_j, \alpha_{j-1}) + d_H(\alpha_j, \alpha_{j-2}) + d_H(\alpha_{j-1}, \alpha_{j-2}) \leq 2m + \delta, \tag{4.5}$$

where $d_H(\alpha_j, \alpha_{j-1})$ is the Hamming distance between the running key bits that correspond to α_j and α_{j-1}, $m = \log M$, and $\delta = 2, 4, \ldots, 2m + 4$. The parameter δ can be chosen as 2 in most cases. Then one can get an irregular full mapping such as

$$\begin{pmatrix} k_j \\ \alpha_{\mathrm{map}(k_j)} \\ \Pi_{\mathrm{map}(k_j)} \end{pmatrix} = \begin{pmatrix} 31 & 10 & 7 & M & \ldots & 16 \\ \alpha_1 & \alpha_2 & \alpha_3 & \alpha_4 & \ldots & \alpha_M \\ 0 & 1 & 0 & 1 & \ldots & 0 \end{pmatrix}. \tag{4.6}$$

For each received qumode, for example, knowing the key k_1, what Bob needs to do is to discern between the two states $x_1 = 0$ or $x_1 = 1$ for the qumode $\left|\alpha_{k_1 + M\left(x_1 + \Pi_{k_1}\right)}\right\rangle$ to determine the data x_1, which is a binary decision problem (see Fig. 4.1). Without knowing the key, Eve on the other hand needs to carry out measurement that can optimally determine the value of the angle for the qumode. The number of qumode pairs M and the mean photon number of the qumodes are chosen in such a way that the quantum noise arises in Eve's measurement makes her unable to determine the true qumode exactly. The mean-square error of such measurement goes as $1/\alpha^2$ for large M. When $M \gg \alpha^2$, Eve's error probability on the data bit tends to 1/2 in an individual attack, which is the purely guessing level.

4.2.2 *Current Experimental Status*

The first experiment of the $\alpha\eta$ protocol for data encryption was implemented using the polarization of mesoscopic coherent states for the qumodes. Bob's receiver operated at 200 kHz, with the average number of received photons $\alpha_0^2 = 27$ and number of qumodes $M = 50$ [15]. It was soon demonstrated for use in a wavelength-division-multiplexed fiber-optic network with amplifiers [5, 19]. These early works were mainly contributed by the inventor of KCQ and collaborators in the U.S.

On the other hand, state-of-the-art high speed Y00 experiments have been carried out by researchers in Japan based on amplitude or intensity modulation [16]. Y00 at 2.5 Gb/s by using 4096-intensity level signals [16, 20] and Y-00 at 10 Gb/s using 64-intensity level signals [21] were demonstrated. Recent works have demonstrated a wavelength-division multiplexing (WDM) transmission of 100-Gbit/s Y00 cipher in a 120-km long optical fiber transmission line [22]. More complex KCQ based on quadrature amplitude modulation (QAM) with data rate as high as 40 Gbit/s over 480 km has also been reported very recently [23, 24]. It should be noted that these numbers refer to the data rate for direct data encryption, not key generation, for rigorous analysis on the key security was not performed.

4.2.3 Comparison Between QKD and KCQ

The KCQ method is based on a fundamentally different concept to protect the classical information than QKD. The following table summarizes the essential differences between QKD and KCQ (Table 4.1).

Table 4.1 The essential differences between QKD and KCQ

	QKD	KCQ
Purpose	Key generation	Currently data encryption (can be used for key generation)
Means of advantage creation	Intrusion-level detection	Asymmetric optimal measurement
Intrusion detection	Precise intrusion-level estimation needed to bound information leak	Not needed but can be implemented to increase security/data rate
Use of pre-shared key	Not required by design (needed for authentication to avoid MIM)	Essential for encryption
Man-in-the-middle (MIM) attack	Prone to	Not prone to (due to the use of pre-shared key)
Mean number of photons	$\sim$0.1 (non CV-QKD)	$\gg$1 (10–1000 and above)
Max. data rate reported	$\sim$1Mbit/s	>10 Gbit/s
Detector technology	Single-photon detector	Conventional photo-detector
Long distance application	Quantum repeater	Conventional optical amplification

4.3 Security Analysis of KCQ

Since a pre-shared key is used, KCQ is susceptible to ciphertext-only attacks (CTA) on the data (plaintext X_n) and the key K (seed key for the case of a cipher using PRNG), and known-plaintext attack (KPA) on the key, where n is the length of the data. For CTA, Eve attempts to find the plaintext X_n and key K by the ciphertext Y_n only. For KPA, Eve attempts to find the key by many pairs of known plaintext and the corresponding ciphertext. In the security analysis of KCQ, we use the threat model with the assumptions that

(i) unlike QKD, a full copy of the quantum state is granted to Eve, so there is no need for quantitative intrusion-level estimation to ascertain her information as a function of her disturbance;
(ii) no sufficient quantum memory is available to Eve to store the quantum states for a duration long enough before she could know of the key K (if ever) in the case of KPA analysis. Thus, Eve is required to make measurements before learning the key in order to gain any information.

4.3.1 *Information-Theoretic (IT) Security*

IT security on the data means that Eve cannot determine uniquely the plaintext from the ciphertext, i.e., $H(X_n|Y_n) \neq 0$, even with unlimited computational power. Here $H(X|Y)$ is the conditional Shannon entropy. For a conventional cipher, Shannon showed that $H(X_n|Y_n) \leq H(K)$, which is called the Shannon limit [25]. Perfect security means that the plaintext is statistically independent of the ciphertext, i.e., $H(X_n|Y_n) = H(X_n)$. Hence perfect security can be attained by a conventional cipher only if $H(X_n) \leq H(K)$, i.e., the key needs to be as long as the plaintext that refers to the one-time pad.

With a conventional cipher, there is no way to exceed the Shannon limit. Nevertheless, exceeding the Shannon limit can lead to secure fresh key generation. Precisely, it was shown that $H(X_n|Y_n^E) > H(K)$ implies $I(X_n; Y_n^E K) < H(X_n)$ with $I(X; Y)$ being the mutual information, a condition that allows key generation [3, 14]. This can be equivalently written as,

$$\Delta I = I(X_n; Y_n^B K) - I(X_n; Y_n^E K) > 0, \tag{4.7}$$

which is the generalization of the condition for the information-theoretic existence proof of key generation for the wiretap channel [4, 8, 26] to include the use of a shared secret key K. Here Y_n^B (Y_n^E) is from Bob's (Eve's) measurement. ΔI is known as the secrecy capacity when optimized with respect to all choices of the distribution for the plaintext. It is related to the fresh key generation rate.

Currently most of the proofs of the IT security of KCQ give the lower bounds or asymptotic bounds only [3, 14, 27–30]. In particular, it is known that the binary detection version of KCQ (Y00 protocol) is not IT-secure under CTA and KPA for large enough n, because the pseudo-random number generated by K (using a linear feedback shift register (LFSR) as an example) has a finite periodic output of period $2^{|K|}-1$ and hence Eve's probability of success to estimate the key goes to unity as the number of periods becomes infinite [27]. Nevertheless, $2^{|K|}-1$ is a huge number for $|K|\geq 128$ and practically the PRNG embedded into KCQ is not to be used longer than its period set by the register length as in the case of standard ciphers. The practical question thus is: what is the maximal secrecy capacity given the seed key with length $|K|$, data length $n<\left(2^{|K|}-1\right)/\log M$, coherent states with mean photon number α_0^2, and the number of qumode sets M? It is estimated that $\Delta I\sim H(K)$ but no rigorous proof has been obtained so far.

In [3, 14], the KCQ inventor argued qualitatively that key generation is possible with the Y00 scheme by giving the bounds of the error rates for Bob's and Eve's optimal measurements under the individual attack. Specifically, when the data length n is smaller than $<\left(2^{|K|}-1\right)/\log M$, the key generation can be estimated from the performance bound obtained by granting Eve the value of the seed key K *after* her individual qumode measurements. In this case, Eve could use the key value to solve the binary decision problem just as what Bob does. Since Bob solves the binary decision problem with the key, he could use an optimal binary quantum receiver to determine the data bit of the qumode. The receiver essentially is to discriminate two equally likely coherent states $\{|\alpha_0\rangle,|-\alpha_0\rangle\}$, no matter what the angle θ_i is in Eq. (4.1). In this case, the error rate of such a receiver is given by

$$P_{\text{Bob}}=\frac{1}{4}e^{-4\alpha_0^2}. \tag{4.8}$$

On the other hand, since Eve makes the decision after she makes her measurements, one can take that her best measurement options would be heterodyne measurements or phase measurements for even better performance, though there is no known physical realization for the latter. The error rate of those receivers are approximately given by

$$P_{\text{Eve}}^{(\text{het})}\sim\frac{1}{2}e^{-\alpha_0^2}, \tag{4.9}$$

$$P_{\text{Eve}}^{(\text{ph})}\sim\frac{1}{2}e^{-2\alpha_0^2}. \tag{4.10}$$

Therefore, for a mesoscopic signal level with $\alpha_0^2=6$, we obtain $P_{\text{Bob}}\sim 10^{-11}$, $P_{\text{Eve}}^{(\text{het})}\sim 10^{-3}$, and $P_{\text{Eve}}^{(\text{ph})}\sim 10^{-6}$. Thus if the data rate is 1 Gbps, Bob is likely to get all 10^9 bits error free per second, whereas Eve's error would be 10^3–10^6 bits within the 10^9 bits depending on where a phase receiver or a heterodyne receiver is used. As a result, Bob in principle can use privacy amplification to eliminate Eve's information and arrive at a secure key of $\sim 10^3$ bits long.

On the other hand, in [29], it was shown more quantitatively that, based on the quantum Gaussian channel model, there in principle exist ciphers exceeding the Shannon limit with the secret capacity given as ΔI for both free space optical (FSO) and optical fiber communications. A wiretap channel Y00 scheme was also proposed to amplify the quantum gain of the legitimate user with respect to the adversary by considering the coding scheme. Yet only partial results were reported. Other quantitative results such as the lower bounds to the average number of spurious keys and the unicity distance under CTA and KPA were also derived [27, 31]. These results nevertheless do not give precise meaning to the practical security (or insecurity) of the KCQ scheme because the actual complexity of key determination as a function of data length is not known.

4.3.2 Complexity-Theoretic (CT) Security

The KCQ scheme (especially the Y00 implementation) is also susceptible to exhaustive search of the key. For the Y00 protocol, the number of qumode basis pairs that are concealed by quantum noise (the so-called masking factor, denoted by Γ) in Eve's measurement can be less than the total number of pairs M. Under the wedge-approximation, for KPA on the key, such masking effect has an additional brute force search complexity given by $\Gamma^{|K|/\log M}$ even when $H\left(K|X_n Y_n^E\right) = 0$, compared to a conventional cipher [3, 30]. The success probability of estimation of the key by Eve in KPA is given by

$$P(K) = \Gamma^{-|K|/\log M} \rightarrow 2^{-|K|(1-1/\log M)}, \tag{4.11}$$

with $\Gamma = M/2$ for an ideal Y00 with proper randomization through the mapping function. This provides security against KPA for a very long observed sequence which cannot be obtained by a mathematical cipher based on PRNG. For CTA, the Y00 has a corresponding complexity-theoretic security with the ideal Y00 having $\Gamma = M$. A large M should be used according to Eq. (4.5). Note that this complexity security analysis applies to individual attacks on the key only.

4.4 Summary

This chapter has introduced a type of keyed communication based on quantum noise (KCQ) to protect the information on the physical layer from an intruder. The method exploits the asymmetry of optimal measurements with and without the pre-shared key. The protocol is able to circumvent many of the challenges of QKD protocols. In particular, many more photons per pulse can be used in KCQ and the detector for the legitimate user can simply be an ordinary photo-diode. In addition, many of the conventional telecommunication techniques can be applied to KCQ to improve the security and transmission data rate.

Despite its nice features for practical quantum communication, the many new properties of KCQ have not been thoroughly explored, even though plug-and-play systems are already available [32]. A rigorous information theoretic security analysis with tight bounds to the security performance is still needed.

References

1. Gisin, N., Ribordy, G., Tittel, W., & Zbinden, H. (2002). Quantum cryptography. *Reviews of Modern Physics, 74,* 145.
2. Lo, H.-K., Curty, M., & Tamaki, K. (2014). Secure quantum key distribution. *Nature Photonics, 8,* 595.
3. Yuen, H. P. KCQ: A new approach to quantum cryptography I. General principles and key generation. http://arxiv.org/abs/quant-ph/0311061v6.
4. Wyner, A. D. (1975). The wire-tap channel. *Bell System Technical Journal, 54,* 1355.
5. Corndorf, E., Liang, C., Kanter, G. S., Kumar, P., & Yuen, H. P. (2005). Quantum-noise randomized data encryption for wavelength-division-multiplexed fiber-optic networks. *Physical Review A, 71,* 062326.
6. Chan, K. W. C., El Rifai, M., Verma, P. K., Kak, S., & Chen, Y. (2015). Security analysis of the multi-photon three-stage quantum key distribution. *International Journal on Cryptography and Information Security (IJCIS), 5*(3/4), 1–13.
7. El Rifai, M., Chan, K. W. C., & Verma, P. K. (2015). Multi-stage quantum secure communication using polarization hopping. *Security Communication Networks, 8,* 4333.
8. Maurer, U. (1993). Secret key agreement by public discussion from common information. *IEEE Transactions on Information Theory, 39,* 733.
9. Kish, L. B. (2006). Totally secure classical communication utilizing Johnson(-like) noise and Kirchoff's law. *Physics Letters A, 352,* 178.
10. Zhuang, Q., Zhang, Z., Dove, J., Wong, F. N. C., & Shapiro, J. H. (2016). Floodlight quantum key distribution: A practical route to gigabit-per-second secret-key rates. *Physical Review A, 94,* 012322.
11. Saad, W., Zhou, X., Debbah, M., & Poor, H. V. (2015). Wireless physical layer security: Part 1. *IEEE Communications Magazine, 53,* 15.
12. Baldi, M., & Tomasin, S. (2016). *Physical and data-link security techniques for future communication systems.* Springer.
13. Wang, H.-M., & Zheng, T.-X. (2016). *Physical layer security in random cellular networks.* Springer.
14. Yuen, H. P. (2009). Key generation: Foundations and a new quantum approach. *IEEE Journal on Selected Topics Quantum Electronics, 15,* 1630.
15. Barbosa, G. A., Corndorf, E., Kumar, P., & Yuen, H. P. (2003). Secure communication using mesoscopic coherent states. *Physical Review Letters, 90,* 227901.
16. Hirota, O., Sohma, M., Fuse, M., & Kato, K. (2005). Quantum stream cipher by the Yuen 2000 protocol: Design and experiment by an intensity-modulation scheme. *Physical Review A, 72,* 022335.
17. Donnet, S., Thangaraj, A., Bloch, M., Cussey, J., Merolla, J. M., & Larger, L. (2006). Security of Y-00 under heterodyne measurement and fast correlation attack. *Physics Letters A, 356,* 406.

18. Shimizu, T., Hirota, O., & Nagasako, Y. (2008). Running key mapping in a quantum stream cipher by the Yuen 2000 protocol. *Physical Review A, 77,* 034305.
19. Liang, C., Kanter, G. S., Corndorf, E., & Kumar, P. (2005). Quantum noise protected data encryption in a WDM network. *IEEE Photonic Technology Letters, 17,* 1573.
20. Harasawa, K., Hirota, O., Yamashita, K., Honda, M., Ohhata, K., Akutsu, S., et al. (2011). Quantum encryption communication over a 192-km 2.5-Gbit/s line with optical transceivers employing Yuen-2000 protocol based on intensity modulation. *Journal of Lightwave Technology, 29*(3), 323–361.
21. Doi, Y., Akutsu, S., Honda, M., Harasawa, K., Hirota, O., Kawanishi, S., Ohhata, K., Yamashita, K. (2010). 360 km field transmission of 10 Gbit/s stream cipher by quantum noise for optical network. In *Proceeding optical fiber communication conference (OFC), OWC4.*
22. Futami, F. (2014). Experimental demonstrations of Y-00 cipher for high capacity and secure optical fiber communications. *Quantum Information Processing, 13,* 2277.
23. Nakazawa, M., Yoshida, M., Hirooka, T., & Kasai, K. (2014). QAM quantum stream cipher using digital coherent optical transmission. *Optics Express, 22,* 4098.
24. Yoshida, M., Hirooka, T., Kasai, K., & Nakazawa, M. (2016). Single-channel 40 Gbit/s digital coherent QAM quantum noise stream cipher transmission over 480 km. *Optics Express, 24,* 652.
25. Shannon, C. (1949). Communication theory of secrecy systems. *Bell System Technical Journal, 28,* 656.
26. Csiszár, I., & Körner, J. (1978). Broadcast channels with confidential messages. *IEEE Transactions on Information Theory, 24,* 339.
27. Nair, R., & Yuen, H. P. (2008). Comment on: "Exposed-key weakness of αη" [Phys. Lett. A 370 (2007) 131]. *Physics Letters A, 372,* 7091.
28. Mihaljević, M. J. (2007). Generic framework for the secure Yuen 2000 quantum-encryption protocol employing the wire-tap channel approach. *Physical Review A, 75,* 052334.
29. Hirota, O., & Sohma, M. (2011). Towards a new way of quantum communication: Getting around the shannon limit of cryptography. *Tamagawa University Quantum ICT Research Institute Bulletin, 1*(1), 1–13.
30. Hirota, O. (2007). Practical security analysis of a quantum stream cipher by the Yuen 2000 protocol. *Physical Review A, 76,* 032307.
31. Nair, R., Yuen, H. P., Corndorf, E., Eguchi, T., & Kumar, P. (2006). Quantum-noise randomized ciphers. *Physical Review A, 74,* 052309.
32. Futami, F., Tanizawa, K., Kato, K., Hirota, O. (2017). Experimental investigation of security parameters of Y-00 quantum stream cipher transceiver with randomization technique, Part I. In *Proceedings volume 10409, quantum communications and quantum imaging XV; 1040901.*

Chapter 5
The Three-Stage Protocol: Its Operation and Implementation

This chapter introduces the three-stage multi-photon protocol, its operation and implementation in a laboratory environment. The implementation uses free-space optics as the transmission medium. Parts of this chapter are based on the authors' work previously reported in [1].

5.1 Introduction

The Three-stage Quantum Cryptography protocol was first proposed in [2]. In this protocol, each party uses its own secret key. The security of the three-stage protocol is based on the fact that Alice and Bob use their own secret keys to transfer secure information to the other party. In the BB84 protocol, each transmitted qubit can be in one of four different states. However, in the three-stage protocol, polarization of the information bits transmitted can be in an arbitrary state. This chapter offers an experimental proof of concept of the three-stage protocol over Free Space Optics (FSO). The transformations needed to implement the three-stage protocol are discussed in detail.

5.2 Principle of Operation

The three-stage protocol shown in Fig. 5.1 was first proposed in [2], and implemented as a multi-photon tolerant protocol in [3]. It presents an alternative to the wide arsenal of cryptographic protocols, including conventional and emerging protocols, such as the BB84. The latter protocol is used to transfer secure keys between communicating parties.

The mode of operation of the three-stage protocol can be described as follows: a sender, Alice, wants to convey a message X to a receiver, Bob. Alice encodes each

P. K. Verma et al., *Multi-photon Quantum Secure Communication*,
Signals and Communication Technology,
https://doi.org/10.1007/978-981-10-8618-2_5

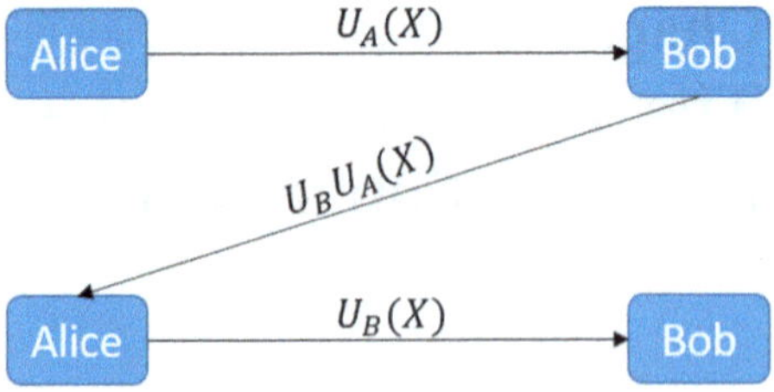

Fig. 5.1 Three-stage protocol operation

bit of her message into a photonic burst consisting of multiple photons. As an example, she encodes bit 0 with 0° polarization and bit 1 with 90° polarization. The two polarization states are orthogonal to each other. Other than orthogonality, there is no restriction on the states of polarization associated with the information bits 0 and 1. In order to secure the information transfer over the channel, Alice and Bob apply secret unitary transformations, U_A and $U_{B,}$ respectively. These transformations should commute, i.e., $U_AU_B = U_BU_A$. The steps of the three-stage protocol are as follows:

Step 1: Alice applies a unitary transformation U_A on the photons encoded with information X and sends them to Bob.

Step 2: Bob Applies $U_B(X)$ on the received photons $U_A(X)$, giving $U_BU_A(X)$ and sends them back to Alice. As mentioned above, U_A and U_B should be commutative transformations. The choice of transformation by Bob is independent of the transformation Alice applied at her end.

Step 3: Alice Applies $U_A^\dagger$ (transpose complex conjugate of U_A) on the received photons to get $U_A^\dagger U_BU_A(X) = U_B(X)$ and sends them back to Bob.

Step 4: Then Bob applies $U_B^\dagger$ on $U_B(X)$ to get the information X.

The operation of the three-stage protocol is shown in Fig. 5.1.

The transformation of the key or data sent by either Alice or Bob is effected by using a rotational change in polarization. Alice and Bob can use any secret transformation they are capable of generating that follows the commutative property, i.e., if U_A and U_B are Alice's and Bob's secret transformations, then,

$$U_AU_B = U_BU_A, \tag{5.1}$$

for all values of U_A and U_B used for communication. These are the basic properties on which the successful operation of the protocol relies. One example of the secret transformation, in terms of the Jones matrix, is a polarization rotation over the plane of linear polarizations given by:

$$U_A = R(\theta) = \begin{bmatrix} \cos\theta & -\sin\theta \\ \sin\theta & \cos\theta \end{bmatrix} \quad \text{and,} \tag{5.2}$$

$$U_B = R(\phi) = \begin{bmatrix} \cos\phi & -\sin\phi \\ \sin\phi & \cos\phi \end{bmatrix}, \tag{5.3}$$

where the Jones vector is defined in the basis of horizontal and vertical polarizations. This rotation operator would change the polarization through an angle of θ or φ (explained in Chap. 2) but not the relative phase between the electric fields of the two polarization components. We can understand this concept better with the help of Stokes' parameters [4]. Any change in polarization due to the rotation operator over the plane of linear polarizations will affect only the parameters, S_1 and S_2. The Stokes' parameter S_3 will remain unchanged. The rotation operator satisfies the commutative property for any combination of θ and φ. The relevance of the commutative operator can be understood through the operation of the three-stage protocol.

Another form of rotation operator, which is also known as a complex rotation operator, is given by,

$$U_A(\theta) = \frac{1}{\sqrt{2}} \begin{bmatrix} e^{i\theta} & e^{-i\theta} \\ ie^{i\theta} & -ie^{-i\theta} \end{bmatrix}, \quad \theta \in [0, 2\pi] \tag{5.4}$$

which can be written as,

$$U_A(\theta) = \begin{bmatrix} \cos\theta & \sin\theta \\ -\sin\theta & \cos\theta \end{bmatrix} \cdot \begin{bmatrix} 1 & 0 \\ 0 & i \end{bmatrix} \cdot \frac{1}{\sqrt{2}} \begin{bmatrix} 1 & 1 \\ 1 & -1 \end{bmatrix}, \quad \theta \in [0, 2\pi] \tag{5.5}$$

In this case, the rotation is out of the plane of linear polarizations. For this operator, all three Stokes' parameters S_1, S_2 and S_3 are changed.

5.3 Implementation of the Three-Stage Protocol Over Free Space Optics (FSO)

The first implementation of the three-stage protocol depicted in Fig. 5.2 was reported in [3, 5, 6]. The implementation of the protocol was divided into three main stages: encoding, polarization transformation, and decoding. At the first stage, each bit of the message was encoded using one of two orthogonal polarization angles; bit 0 was encoded with 0° polarization and bit 1 was encoded with 90° polarization. At this stage, an optical beam polarized at 45° was directed into a polarizing beam splitter (PBS), where it was divided into two equal intensity beams. The bits to be transferred from Alice to Bob were encoded using LabView programmed shutters. Shutters were programmed in a way such that shutter 1 would open when bit 1 was to be sent; the optical beam would then pass through a 90° polarizer (pol-1). On the other hand, if bit 0 was to be sent, shutter 2 would open and the optical beam would pass through a 0° polarizer (pol-2). After encoding each

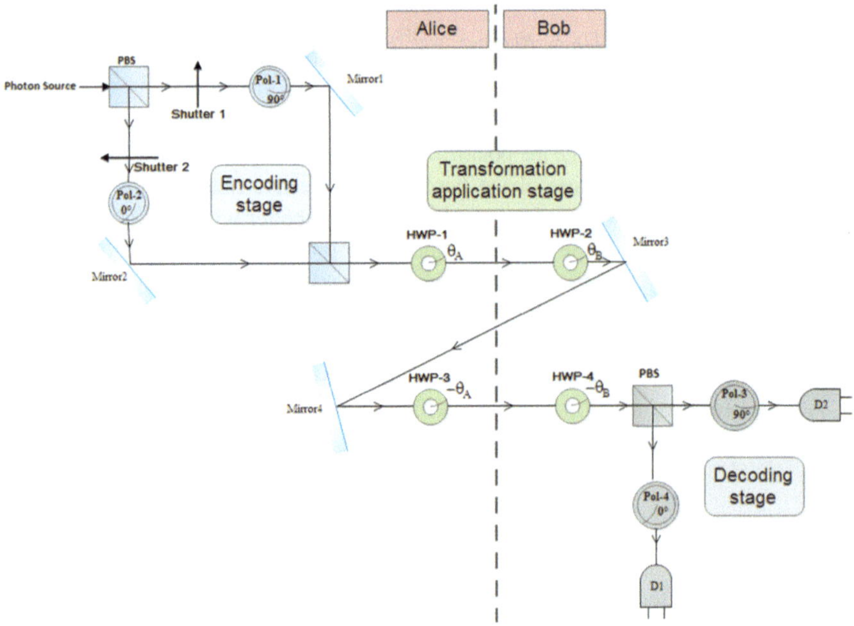

Fig. 5.2 Implementation of the three-stage protocol [5]

bit with the respective polarization, the optical beam was redirected using a mirror (mirror 1 in the first path and mirror 2 in the second path) and a combiner.

During the second stage of the transformations, U_A and U_B were applied using half wave plates. Alice had two half wave plates mounted on rotators controlled by a LabView program. Half wave plate 1 (HWP-1) was rotated to a random angle θ_A (angle of the fast axis) chosen by Alice and known only to her; meanwhile half wave plate 2 (HWP-2) was rotated to an angle $-\theta_A$. At Bob's side two half wave plates were fixed at angles θ_B and $-\theta_B$ (HWP-3 and HWP-4 respectively). It should be noted that the half wave plates at Bob's side were also mounted on rotators and the value of θ_B were randomly chosen and known only to Bob. The formalism of how the angles of the fast axis were chosen is discussed in the next section using the Mueller matrices.

In the last stage of this implementation, the decoding stage, the optical beam passed through a polarization beam splitter. The horizontal polarization component of the optical beam passed the beam splitter, while the vertical component was reflected. Decoding was done according to the light intensity received at the detectors D1 and D2.

The data rates achieved were of the order of a few bits per second due to the fact that the implementation used mechanically driven hardware to encode and effect the polarization rotations. The experiment was implemented with a free space channel distance of 30 cm.

5.3.1 Rotation Transformations

In this section, the setup of the unitary transformations applied using half wave plates is discussed along with the choice of the rotation angle (θ) of the half wave plates with respect to the horizontal axis. This choice is based on the Mueller matrix formalism to ensure that a polarization angle input to the setup of the half wave plate is equal to that at the output of the setup.

5.3.2 Half Wave Plate Operation

A half wave plate produces a polarization shift of 180° between the fast and slow axes of a wave plate. As detailed in Chap. 2, the Mueller matrix of a half wave plate is given by:

$$M_{\mathrm{HWP}} = \begin{bmatrix} 1 & 0 & 0 & 0 \\ 0 & 1 & 0 & 0 \\ 0 & 0 & -1 & 0 \\ 0 & 0 & 0 & -1 \end{bmatrix}. \tag{5.6}$$

The implementation described in the previous section uses rotating half wave plates. The Mueller matrix of a rotating half wave plate with an angle θ with respect to the horizontal direction is given by,

$$M(\theta)_{\mathrm{HWP}} = M_{\mathrm{rot}}(2\theta) \cdot M_{\mathrm{HWP}} \cdot M_{\mathrm{rot}}(-2\theta), \tag{5.7}$$

where $M_{\mathrm{rot}}(\theta)$ is the Mueller matrix for rotation and is given by:

$$M_{\mathrm{rot}}(\theta) = \begin{bmatrix} 1 & 0 & 0 & 0 \\ 0 & \cos\theta & -\sin\theta & 0 \\ 0 & \sin\theta & \cos\theta & 0 \\ 0 & 0 & 0 & 1 \end{bmatrix}. \tag{5.8}$$

Note that θ here is the actual physical rotation angle of the half-wave plate, which effects a polarization rotation by an angle of 2θ. Then the Mueller matrix of a rotating half wave plate is given by:

$$M_{\mathrm{HWP}}(\theta) = \begin{bmatrix} 1 & 0 & 0 & 0 \\ 0 & \cos(4\theta) & \sin(4\theta) & 0 \\ 0 & \sin(4\theta) & -\cos(4\theta) & 0 \\ 0 & 0 & 0 & -1 \end{bmatrix}. \tag{5.9}$$

5.3.2.1 Choice of the Rotation Angle

It is important to note that the requirement imposed on the transformations used in the three-stage protocol is to be commutative while being only known to the entity applying them. In cases where only the setup of Alice's half wave plates is considered, one can see that when Alice applies a transformation $M_{\mathrm{HWP}}(\theta_A)$ using her first half wave plate (HWP-1 in Fig. 5.2), she should apply $M_{\mathrm{HWP}}(\theta_A)$ using her second half wave plate (HWP-3 in Fig. 5.2) in order to remove the effect of her first transformation on the input polarization angle. It can be observed that

$$M_{\mathrm{HWP}}(\theta_A) \cdot M_{\mathrm{HWP}}(\theta_A) = I, \tag{5.10}$$

where I is the identity matrix.

However, once the half wave plates of Bob are considered, one can see that the polarization of the input beam is not equal to the polarization of the output beam even if the same angles are used, viz.,

$$M_{\mathrm{HWP}}(\theta_B) \cdot M_{\mathrm{HWP}}(\theta_A) \cdot M_{\mathrm{HWP}}(\theta_B) \cdot M_{\mathrm{HWP}}(\theta_A) \neq I. \tag{5.11}$$

This is because the operation of the half wave plates is not commutative, i.e.,

$$M_{\mathrm{HWP}}(\theta_B) \cdot M_{\mathrm{HWP}}(\theta_A) \neq M_{\mathrm{HWP}}(\theta_A) \cdot M_{\mathrm{HWP}}(\theta_B). \tag{5.12}$$

One can note from the Mueller matrix representation of half wave plates that

$$M_{\mathrm{HWP}}(\theta_B) \cdot M_{\mathrm{HWP}}(\theta_A) = M_{\mathrm{HWP}}(-\theta_A) \cdot M_{\mathrm{HWP}}(-\theta_B). \tag{5.13}$$

Thus,

$$\begin{aligned} M_{\mathrm{HWP}}(-\theta_B) \cdot M_{\mathrm{HWP}}(-\theta_A) \cdot M_{\mathrm{HWP}}(\theta_B) \cdot M_{\mathrm{HWP}}(\theta_A) &= M_{\mathrm{HWP}}(-\theta_B) \cdot M_{\mathrm{HWP}}(-\theta_B) \\ &\quad \cdot M_{\mathrm{HWP}}(\theta_A) \cdot M_{\mathrm{HWP}}(\theta_A) = I, \end{aligned} \tag{5.14}$$

where $0 \leq \theta_A \leq \pi$ and $0 \leq \theta_B \leq \pi$.

Equation (5.14) shows the choice of angles that will insure that the transformations used by Alice and Bob commute without the need of sharing any information about the actual angles. Therefore, we choose the angles of the transformation in such a way that in case Alice applies $M_{\mathrm{HWP}}(\theta_A)$ first, and then she will apply $M_{\mathrm{HWP}}(-\theta_A)$ to cancel her transformation. Bob does the same using his own randomly chosen angles.

5.4 Summary

This chapter has presented the three-stage protocol. A detailed discussion of its proof of concept implementation over free space optics has been presented. As discussed earlier, the security of the three-stage protocol is based on the usage of rotation transformation at the sender and receiver. These transformations are only known to the party applying them. The rotations application is explained using the formalism of Mueller matrices.

References

1. El Rifai, M., Chan, K. W. C., & Verma, P. K. (2015). Multi-stage quantum secure communication using polarization hopping. *Security and Communication Networks.*
2. Kak, S. (2006). A three-stage quantum cryptography protocol. *Foundations of Physics Letters, 19*(3), 293–296.
3. Mandal, S., et al. (2012). Implementation of secure quantum protocol using multiple photons for communication. arXiv preprint arXiv:1208.6198.
4. Collett, E. (2003). *Polarized light in fiber optics*. Bellingham: SPIE Press.
5. Mandal, S. (2012). *Implementation of the three-stage protocol over free space optics*. Norman: University of Oklahoma.
6. Chen, Y., et al. (2013). Multi-photon tolerant secure quantum communication—From theory to practice. In *2013 IEEE International Conference on Communications (ICC)*. IEEE.

Chapter 6
The Multi-stage Protocol

This chapter generalizes the three-stage protocol into a family of multi-stage protocols. It compares the multi-stage protocol with single-photon protocols and illustrates how a multi-photon protocol can be made secure against man-in-the-middle attack. Since a multi-photon protocol is, in general, subject to photon-siphoning attacks, the protocol introduces another variable to thwart such attacks. Parts of this chapter are based on the authors' work previously reported in [1–3].

6.1 Introduction

Securing information in transit is an increasingly important need of modern society. As discussed in the previous chapters, given sufficient computational power, the majority of encryption techniques in commercial use today can be rendered ineffective. This has led to exploring techniques based on quantum mechanics which offers the only known means to provide unconditional security.

BB84 and its variants [4–10] are deployed for quantum key exchange and are based on single photon implementations which have their own set of drawbacks. These drawbacks were discussed in detail in Chaps. 1 and 2. We note that it is technologically challenging to have a device that can reliably generate single photons with guaranteed periodicity. Thus, contemporary implementations rely on weak optical beams that produce much less than one photon per time slot on average. Therefore, most time slots will be empty; a few would have single photons, and very few more than a single photon. The BB84 configuration thus limits the distance and speed of quantum communication.

Attempts have been made in establishing quantum secure protocols that use multiple photons in order to overcome the limitations associated with approaches based on single photons [11, 12]. This chapter proposes a multi-stage protocol related to the three-stage protocol introduced in 2006 [13–15]. The three-stage protocol was discussed in detail in Chap. 5 of this book. This chapter generalizes

P. K. Verma et al., *Multi-photon Quantum Secure Communication*,
Signals and Communication Technology,
https://doi.org/10.1007/978-981-10-8618-2_6

the three-stage protocol to an m-stage protocol where m defines the number of stages in the protocol. The multi-stage protocol is based on the usage of unitary transformations known only to Alice and Bob, individually. Alice and Bob need not communicate information about the transformations they use to anyone, even to each other. The only condition on these transformations is that they are commutative. In the following, the encoding of the information bits and the transformation operations are performed using the polarization degree of freedom of photons.

In addition, this chapter discusses the Man-in-the middle attack in multi-stage protocols. This chapter also provides two ways to counter the man-in the middle attack. The first technique is at a physical level and can be used if the protocol is implemented over fiber optics using polarization coding. This technique is based on the fact that any attempt to physically intrude on the fiber will result in modification of the behavior of the polarization channel, which can be immediately detected. It is worth noting that, during polarization based implementation of the protocol, Alice and Bob do monitor the channel via a channel characterization procedure. This will be discussed later in this chapter.

The second technique to counter the man-in-the-middle attack is a modified version of the multi-stage protocol [1]. The technique introduces an additional variable in the multi-stage protocol, where the number of variables n is increased by 1, i.e., $n \geq m+1$, m being the number of stages. This modification is called a key/message expansion multi-stage protocol. In this technique, an initial random string of bits known to both Alice and Bob is used to fully secure the next message to be transferred. In addition, the implementation of the key/message expansion protocol using passive optical components for FSO communication along with its Mueller matrices formalism is presented in this chapter.

6.2 The Multi-stage Protocol Polarization Hopping

This section presents the multi-stage protocol, a variation of the three-stage protocol, as a secure protocol that makes use of linearly polarized light to encode classical bits of information to be transferred. During the execution of a multi-stage protocol with an odd number of stages, one of the communicating parties (Alice) starts the communication by encoding the data to be sent into one of two orthogonal states, then applies its unitary transformation and sends the transformed data to Bob. Bob applies his own unitary transformation and sends the resulting state back to Alice. Alice now removes the first unitary transformation she applied, and applies a new unitary transformation before sending it back to Bob, who will repeat the process. This process will be executed in the same way until the last stage of the protocol where Alice will remove all the transformations she made and return the data, with only Bob's transformations, back to the Bob. Since Bob knows exactly the transformations he made to the encoded data, he removes his transformations and recovers the data.

In case m is even, the communication will start at the receiver (Bob). He starts the communication using a random polarization state representing his first unitary transformation. He sends this polarization state to Alice who in turn encodes the information to be shared and applies her transformation. The communication will proceed in this manner as in the case where m is odd, but with Bob being the entity who decodes the information at the last stage. The cases where $m = 1$ or 2 are special cases of the generalized process. For security purposes, the use of the key/message expansion protocol proposed later in this chapter is required.

As explained above, the encoded polarization states of the information to be transferred are never sent directly on the channel. At each stage of transmission, a unitary transformation is applied in order to change the state of the polarization to be sent over the channel. The unitary transformation applied is only known to the party applying it. We call polarization hopping the process of changing the polarization state at each stage of the protocol. The message transfer always starts by encoding a bit of information in a polarization state; for example, bit 0 is encoded using 0° polarization angle and bit 1 is encoded using 90° polarization angle, whereas on the transmission channel the angle polarization is random. A transformation applied by one communicating party at a given stage will result in new values of α and β. Note that in the following we will restrict ourselves to cases with real α and β, i.e., the polarization stays on the equator of the Poincaré sphere.

We designate the polarization state as a value of a linear polarization angle, e.g., 0° is bit 0 and 90° is bit 1. When the transfer is initiated on Alice's side, she performs her unitary transformation by changing the value of the polarization angle and sends it to Bob, who in turn will do the same. The next steps of the protocol will be carried out in the same way until the last stage where Alice removes her transformations and sends them to Bob who in turn removes his transformations and recover the original state. While in transit, the value of the polarization angle will be of the following form:

$$\phi_i = \begin{cases} 0^\circ + \frac{b_i \times 180^\circ}{M} + \frac{a_i \times 180^\circ}{M}, & \text{encoded angle is } 0^\circ \\ 90^\circ + \frac{b_i \times 180^\circ}{M} + \frac{a_i \times 180^\circ}{M}, & \text{encoded angle is } 90^\circ \end{cases}$$

where b_i denotes Bob's transformation and a_i denotes Alice's transformation in the ith stage with $a_i, b_i \in \{0, 1, \ldots, M-1\}$. The total number of angles that can be used over the channel is M, i.e., by dividing the full polarization circle into $2M$ equal parts. Alice's and Bob's transformations are multiples of $\frac{180^\circ}{M}$ changes to the polarization angle. We wish to have M as large as possible for higher degree of security, with the ideal limit $M \to \infty$. Figure 6.1 represents one example of the extraction of possible polarization angles.

The advantage the intended recipient has over an intruder is that, while the intruder has to distinguish between one of the M non-orthogonal polarization angles, the intended recipient only needs to distinguish between two orthogonal polarization angles. This means that while the receiver would need only one photon to do so, the intruder Eve would generally require more than one photon per stage

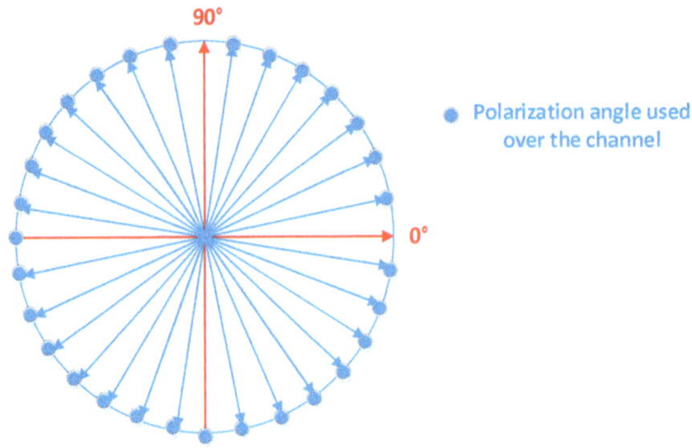

Fig. 6.1 Representation of the choices of encoding angles and the angles used over the channel for $2M = 32$

to accurately determine ϕ_i. In other words, when Eve attempts to measure ϕ_i she will be confronted with quantum noise which she may not be able to overcome with finite probability.

Throughout this chapter, we use the multi-stage protocol, with the number of stages $m = 3$. However, the discussion can be generalized to any value of m. It is important to note that the multi-stage protocols fall into two categories. The first category is the case when $m < 3$. In this case, the protocol needs an initialization vector to be shared before the onset of the transmission, and thus it can only be used as a key/message expansion protocol. The second category is when $m \geq 3$. In this case, the protocol can be used for direct secure communication or can be used as a key/message expansion protocol. It is important to note that an increased number of stages m means that an eavesdropper is faced with the problem of measuring the states of polarization at a correspondingly more number of stages. In addition, using more stages means that the sender can use more photons per pulse to encode each bit of information. However, an increase in the number of stages poses an overhead on the sender and the receiver with a corresponding increase in the number of transmissions required. As stated previously, when m is odd, the communication begins at Alice's side and ends at Bob's side. When m is even, the communication begins and ends both at Bob.

6.2.1 *Comparison with Single-Photon Protocols*

The multi-stage protocol is a multi-photon tolerant protocol that differs from the single-photon protocols in many aspects. We now compare it with the BB84 protocol which is the prime example of the single photon protocol:

(1) Multi-stage protocols use pulses of photonic beams consisting of several photons or more per pulse. The BB84 protocol uses no more than a single photon per pulse.

(2) The underlying physical principles used to prove security are different. The multi-stage protocol security is based on the quantum detection theory or quantum state discrimination by the fact that measuring a state of polarization with fewer numbers of photons than needed will generate quantum noise that an eavesdropper cannot overcome. On the other hand, the BB84 is mainly based on the no-cloning theorem and its underlying fact that an exact copy of a photon cannot be created.
(3) It should be noted that the operation of the multi-stage protocol is comparable to that of the BB84 if the number of photons is restricted to be one. In this case, the multi-stage protocol can operate with only two different polarization transformations at each end of the communication. In other words, the protocol will be using only four different polarization angles to secure the transmission of a single photon.
(4) It might seem that in BB84 no prior secret string (initialization vector) is needed to start the communication. This is not the case since, in order to mutually authenticate the transmitting and the receiving parties, Alice and Bob do need a common secret string to start with. Thus, the requirement that Alice and Bob share some random string (initialization vector) in the key/message expansion multi-stage protocol, is implicitly made in BB84 as well [12].
(5) In the BB84 protocol, the receiver as well as the intruder are required to distinguish among four different polarization states. Thus, both the receiver and the intruder measurements are probabilistic. In the case of a multi-stage protocol, an intruder Eve will be faced with identifying from among an indefinite number of polarization angles. In contrast, the intended recipient only needs to distinguish between two orthogonal polarization angles.

6.3 Man-in-the-Middle Attack

The nomenclature of the Man-In-The-Middle is derived from the basketball situation where a couple of players are trying to pass a ball to each other. However, there is one player between them who is trying to take hold of the ball. The common scenario of a Man-In-The-Middle attack involves two endpoints (victims: Alice and Bob), referred to as sender and receiver, and a third party (attacker: Eve). The attacker can gain access to the communication channel between two endpoints, and can siphon off and manipulate their messages. The Man-In-The-Middle attack can be visualized in Fig. 6.2. In Fig. 6.2, the intruder Eve impersonates Bob as far as communication with Alice is concerned. She also disguises herself and appears as Alice as far as communication with Bob is concerned. As a result, Alice and Bob believe they are communicating with each other while, in reality, Eve intercepts the communication and appears as the legitimate end point to both Alice and Bob.

Even though practical ways of authentication exist in practice, Eve can carry out a man-in-the-middle attack on a communication channel. Literature contains

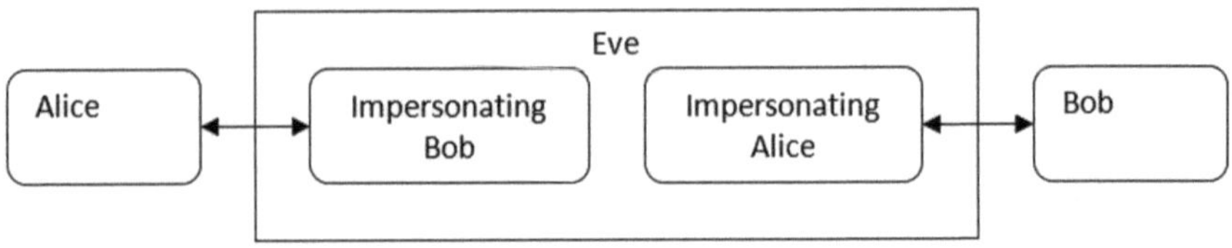

Fig. 6.2 Man-in-the-middle attack

discussions on carrying out a man-in-the-middle attack on the BB84 protocol [16, 17], and on various methods to detect or counter it [18, 19]. A procedure by which Eve can attempt a man-in-the-middle attack on the three-stage protocol has been presented in [20].

The following scenarios describe how the attack is carried out.

1. Eve impersonates Bob and receives a message from Alice. After decoding the message, she regenerates the message, and communicates it to Bob. Bob receives the message sent by Eve but believes it has come from Alice because Eve, in this communication, has impersonated Alice.
2. Eve, like in scenario 1, impersonates Bob and intercepts a secret message from Alice. But now, instead of sending the original message received from Alice, Eve sends a different message to Bob. In this case, only Eve has received the secret message from Alice. Bob receives the message created but Eve but believes it to be a secret message from Alice.
3. Eve can impersonate Bob, but does not decode the message sent by Alice or was not able to decode the message from Alice. However, she impersonates Alice and sends her own message that she wants to convey to Bob, as if the message has come from Alice. In this case, a communication link between Alice and Bob has been completely blocked, and Bob gets a message from Eve, which he thinks is a message from Alice.

In all the three scenarios, Eve was successful in either retrieving a secret message, conveying her own message, or blocking a communication channel completely. The vulnerability of the three-stage protocol to man-in-the-middle attack very clearly demonstrated in [20]. In this section, we propose a procedure by which the man-in-the-middle attack by Eve can be detected. Using this technique, the three-stage protocol implementation over fiber optics can be secured against a man-in-the-middle attack.

Man-in-the-middle Attack

The implementation procedure of the three-stage protocol over optical fiber was discussed in detail in [21]. This implementation consists of a channel characterization procedure. This procedure allows Alice and Bob to model the behavior of the optical fiber channel they are using for their communication. Channel characterization results in determining the value by which the angle of the linear polarization shifts as the optical beam travels from the point it's launched on to the optical fiber to its destination.

In general, channel characterization is carried out over a prolonged period of time so that there is enough confidence on the ranges of variation in the parameters of the channel due to ambient conditions. Knowing the polarization behavior of the optical fiber as a function of environmental factors is important in ensuring that Alice and Bob correctly interpret the data they receive.

The optical channel can be characterized as often as necessary. The frequency of characterization is determined by the variations in the polarization behavior of the optical fiber over time. For example, if the diurnal variations in the ambient temperature are 40 °F, Alice and Bob might choose to increase the frequency of the channel characterization procedure compared to the case when the diurnal temperature variation was merely 15 °F.

Similarly, Alice and Bob might come across a situation when there is an increase in noise arising from vibration caused by, for instance, road repair or repair to the conduit of the fiber cable. This will require more frequent characterization of the polarization channel of the optical fiber.

Alice and Bob can store channel characterization data as a function of time, season, or such other environmental factors that are deemed necessary. Depending upon the applicable environmental conditions, they can use the corresponding channel characterization data so that the angle of correction can be more accurately predicted. A channel characterization frequency that matches the environmental conditions will ensure that the needed correction to recover the data changes smoothly over time rather than experiencing sharp variations over a short period of time.

The storage of this data will allow Alice and Bob to detect any intrusion attempt on the fiber and any of its channels including active man-in-the-middle attack on a channel. To understand the strength of the protocol in addressing a potential man-in-the-middle attack, consider the two plots in Fig. 6.3. Figure 6.3 represents a plot of channel characterization correction angle against time or iteration number. In Fig. 6.3a, there is a smooth curve representing smoothly changing channel characterization angles. In Fig. 6.3b, there is an abrupt change in the channel characterization angle after the sixth iteration. There might, possibly, be multiple reasons for this drastic change of characterization angle, e.g., an abrupt physical change, unexpected bend or twist over the length of the optical fiber, etc. However, one reason could be a man-in-the-middle attack on the channel. If an abrupt change is detected, its cause needs to be identified before a man-in-the-middle attack is eliminated.

Figure 6.4 shows different locations on the optical fiber, identified as black dots, where Eve tries to carry out man-in-the-middle attacks and impersonate Alice and Bob. Eve needs to carry out the channel characterization procedure at the point of attack in order to know the change of polarization angle caused by a particular length of the optical fiber channel. Without knowing this angle, Eve will not be able to retrieve any data or polarization states transmitted by Alice or Bob. Therefore, the channel characterization procedure is a requirement for Eve if she is going to be successful in her attack.

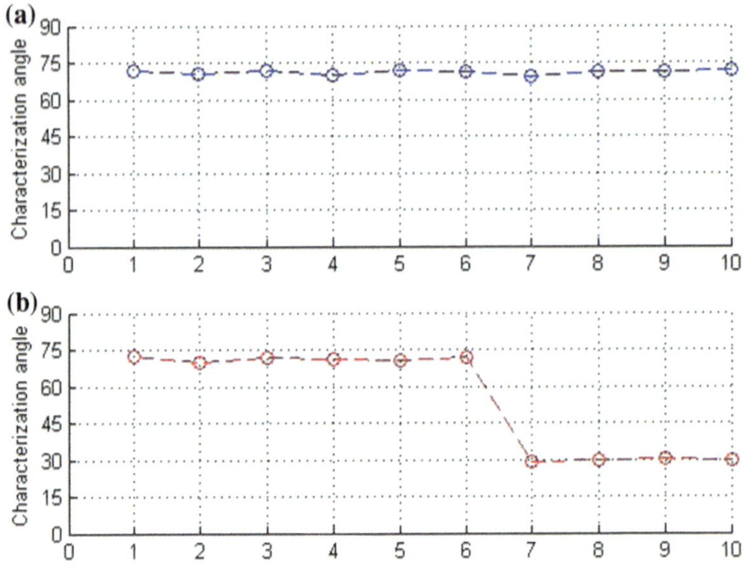

Fig. 6.3 Channel characterization angle iteration outcome

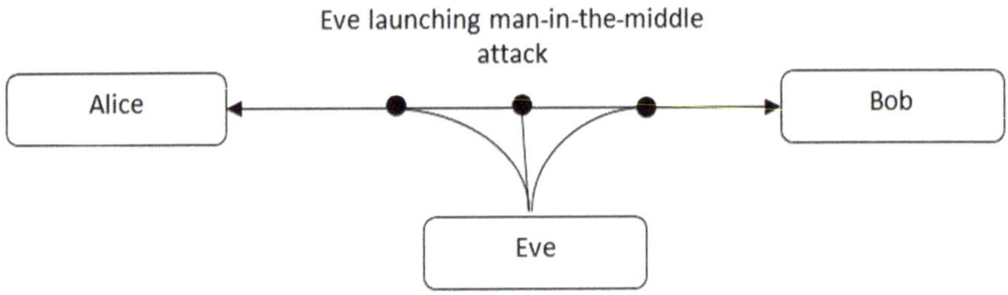

Fig. 6.4 Different locations on the optical fiber, where Eve tries to carry out man-in-the-middle attacks and impersonate Alice and Bob

There is a small probability that Eve will be lucky and carry out her man-in-the-middle attack at a particular location on the channel that corresponds to the characterization angle at Alice's and Bob's locations. However, even in this case, Eve will likely drastically change the characterization angle for Bob in the process of characterizing the channel for her to be able to decipher the message sent by Alice. Thus, Alice and/or Bob can identify the breach and launch an authentication procedure. The parties will then stop sending data to each other and shift their communication to a different optical path obviating the risk of man-in-the-middle attack.

6.4 Key/Message Expansion Multi-stage Protocol

In this section, a key/message expansion algorithm using a multi-stage protocol is discussed. The key/message expansion protocol is a way to counteract a man-in-the-middle attack. We call a key expansion algorithm a key distribution method that requires having a shared initialization vector at the onset of the communication. A message expansion algorithm is a message sharing scheme that requires having a shared initialization vector at the onset of the communication. The initialization vector used is updated using the message or the key shared in the case of message expansion or key expansion, respectively.

6.4.1 *Multi-stage Protocol Using an Initialization Vector*

In this section, we present the multi-stage protocol using $m+1$ variables, where m is the number of stages. It is well known that it is impossible to obtain a unique solution for a system with $m+1$ variables, but only m data points. In the proposed approach, we make sure our system satisfies this condition by adding the requirement of an initialization vector (IV) to any version of the multi-stage protocol. The IV is shared between the sender and receiver at the onset of the communication. The IV, along with the message/key shared, will be used in order to create a new IV to be used during the next communication.

In the following section, we discuss the four-variable three-stage protocol. It should be noted that the same logic can be used to generalize the proposed approach for different values of m. As mentioned earlier, if m is even the communication would start and end at the receiver's end, whereas, if m is odd, the communication starts at the sender and ends at the receiver.

The key/message expansion protocol can be used to counter a man-in-the-middle attack. The usage of an initialization vector known only to the sender and receiver makes it impossible for a third party, Eve, to retrieve any information off the channel without having prior knowledge of the IV used.

6.4.2 *Operation of the Four-Variables Three-Stage Protocol*

In this section, we propose an implementation of the four-variable three-stage protocol reported in [1]. The four-variable three-stage protocol adds one more dimension to the three-stage protocol to enhance its security while keeping its mode of operation intact. The dimension added is presented as an initialization vector IV that Alice and Bob are assumed to possess at the beginning of the first iteration of the protocol. IV is a string of unitary transformations; we call a cycle of the protocol a complete execution of the three-stage protocol that results in one bit shared between Alice and Bob. It is important to note that the number of cycles per iteration is equal to the length of the string IV. IV is updated to a new value at the end of each iteration using the message shared between Alice and Bob as well as the initialization vector used in the last iteration. This can be observed as chaining between iterations of the protocol implementation. One can use any non-linear function to relate between IV_0 and X_0 and generate the new value IV_1. The operation of the four-variable three-stage protocol is discussed below and depicted in Fig. 6.5.

We call $IV_0[n]$ the initialization vector at iteration 0, cycle number n; $X_0[n]$ is the bit value of the message being transferred, and $\alpha_0[n]$ and $\beta_0[n]$ are the values of the unitary transformations at iteration 0 cycle number n of Alice and Bob respectively.

Step 1: Alice applies a unitary transformation $\alpha_0[n]$ on the on $IV_0[n]$ and sends the optical beam to Bob.

Step 2: Bob applies $\beta_0[n]$ on the received optical beam and sends it back to Alice.

Step 3: Alice applies $\alpha_0^{\dagger}[n]$ (transpose complex conjugate of $\alpha_0[n]$) on the received qubit to get and then encodes the value of $X_0[n]$ and sends it to Bob.

Step 4: Then Bob applies $IV_0^{\dagger}[n]$ (transpose complex conjugate of $IV_0[n]$) then applies $\beta_0^{\dagger}[n]$ and gets the information $X_0[n]$.

At the next cycle, Alice will use a new transformation set $\alpha_0[n+1]$ and Bob $\beta_0[n+1]$, along with the next value in the string of the initialization vector $IV_0[n+1]$, When the number of cycles is equal to the length of IV, a new IV_1 of the same length as IV_0 will be generated. It should be noted that Alice and Bob do not have any restrictions on the transformations associated with IV, and it does not need to commute with α and β. Furthermore, the updated initialization vector will be in the binary form containing a sequence of 0's and 1's. Alice and Bob can associate these bits with variable transformation values depending on a prior agreement.

Comparing the first stage of the protocol in Fig. 6.5 with the second one we see that a random variable $\beta[n]$ is added at the second stage. At the third stage $\alpha[n]$ is

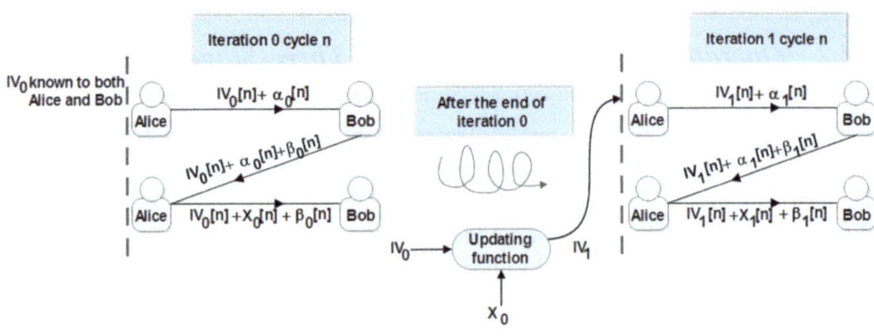

Fig. 6.5 Operation of the three-stage protocol using four variables

removed but $X[n]$ is added. Do note, however, that the consecutive eavesdropper having simultaneous access to the three stages of the protocol will not be able to compute the value of the sent bit, since he/she will be faced with the problem of solving an indeterminate system of equations. In addition, such an approach makes it impossible to launch a man-in-the-middle attack.

The addition of an initialization vector to the three-stage protocol can also be regarded as a door function to protect the message sent over the channel. Any illegitimate user is denied the ability of retrieving the value of the bit sent over the channel as long as he/she does not have in his or her procession the key for the door function.

6.4.3 *Implementation of the Four-Variables Three-Stage Protocol*

In this section, we discuss the implementation of a four-variable three-stage protocol over free-space optics (FSO) using passive optical components. The setup of the implementation is depicted in Fig. 6.6. In the setup, Alice has four half wave plates at her end, and Bob three. At the beginning of the protocol, Alice generates a state with a 0° linear polarization using a 0° polarizer. Then she applies a transformation $IV[n]$ using a half wave plate (HWP-1) set at an angle θ, and another unitary transformation $\alpha[n]$ using her second half wave plate (HWP-2) at an angle θ_A. Then she sends the optical signal to Bob. Bob applies his unitary transformation $\beta[n]$ using his first half wave plate (HWP-3) at an angle θ_B, and sends the optical

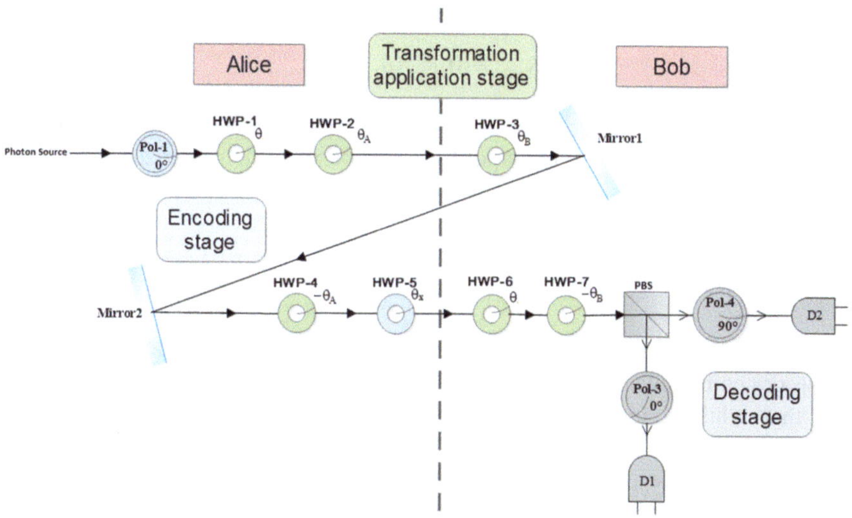

Fig. 6.6 Implementation of the four variables three-stage protocol

beam back to Alice using mirror 1. Alice removes her transformation by setting her third half wave plate (HWP-4) at the angle $\alpha[n] = -\theta_A$. Then she applies the transformation associated with the encoded bit using her fourth half wave plate (HWP-5) set at angle $\theta_x = 0°$ in the case of bit 0 is being sent, and $\theta_x = 45°$ if bit 1 is being sent. At this point, Alice sends the optical beam containing the information back to Bob who first passes it through a half wave plate (HWP-6) set at angle θ to remove the transformation induced by the initialization vector. Then he removes his own transformation using a half wave plate (HWP-7) set at the angle $-\theta_B$. After these operations, Bob has a beam polarized at either 0° or 90°. He will pass it through a polarization beam splitter to detect whether bit 0 or 1 has been received.

The choice of the angles for the setup described above is made according to the same formalism described in Sect. 3.3 of Chap. 5. For the implementation of the four-variable three-stage protocol we can write the Mueller matrices of the half wave plates operation as follows:

$$\begin{aligned} &M_{\text{HWP}}(-\theta_B) \cdot M_{\text{HWP}}(\theta) \cdot M_{\text{HWP}}(\theta_x) \cdot M_{\text{HWP}}(-\theta_A) \cdot M_{\text{HWP}}(\theta_B) \cdot M_{\text{HWP}}(\theta_A) \\ &\quad \cdot M_{\text{HWP}}(\theta) \\ &\quad = M_{\text{HWP}}(\theta_x). \end{aligned}$$

Since the input state is a 0° polarized state, it can be represented using the Stokes parameters:

$$S_{\text{in}} = \begin{bmatrix} 1 \\ 1 \\ 0 \\ 0 \end{bmatrix}.$$

The Stokes' parameters of the output of the proposed setup are given by:

$$S_{\text{out}} = M_{\text{HWP}}(\theta_x) \cdot S_{\text{in}}. \tag{6.1}$$

This means that in the case $\theta_x = 0°$, the output light will be horizontally polarized and if $\theta_x = 45°$, the output light will be vertically polarized.

6.5 Summary

This chapter has proposed a generalized multi-stage multi-photon tolerant protocol for secure communication. The security of the multi-stage protocol is based on the fact that while a legitimate receiver only needs to distinguish between two orthogonal polarization states, an intruder has to distinguish among an infinite number of possible polarization states. In other words, while the receiver would need only one photon to do so, an intruder would need to siphon off several photons per stage.

This chapter has also proposed two techniques to counter a man-in-the-middle attack launched on a multi-stage protocol system. The first technique can be used if the implementation was performed over fiber optics using polarization encoding. This technique uses the fact that an intruder on the communication channel will introduce a change in the behavior of the channel as far as its polarization footprint is concerned. This change can be detected by Alice and Bob and, in turn, the presence of the intruder will be revealed.

The second technique proposed is a key/message expansion scheme associated with the multi-stage protocol. The key/message expansion scheme provides a countermeasure to any man-in-the-middle attack that can be launched on the system. In addition, this chapter has demonstrated an implementation of the key/message expansion scheme. This implementation has been done in a laboratory setup using passive optical components.

References

1. El Rifai, M., & Verma, P. K. (2013). An algorithmic approach to securing the three-stage quantum cryptography protocol. In *2013 12th IEEE International Conference on Trust, Security and Privacy in Computing and Communications (TrustCom)* 2013 (pp. 1803–1807).
2. Punekar, N. V. (2013). *Implementation of the three-stage protocol over optical fiber* (Masters thesis). University of oklahoma Tulsa.
3. El Rifai, M., Chan, K. W. C., & Verma, P. K. (2015). *Multi-stage quantum secure communication using polarization hoppng*. Security and Communication Networks.
4. Bennett, C. H. (1992). Quantum cryptography using any two nonorthogonal states. *Physical Review Letters, 68*(21), 3121.
5. Jin, D., Verma, P. K., & Kartalopoulos, S. V. (2009). Fast convergent key distribution algorithms using a dual quantum channel. *Security and Communication Networks, 2*(6), 519–530.
6. Parakh, A. (2013). A probabilistic quantum key transfer protocol. *Security and Communication Networks, 6*(11), 1389–1395.
7. Zamani, F., & Verma, P. K. (2011). A QKD protocol with a two-way quantum channel. In *2011 IEEE 5th International Conference on Advanced Networks and Telecommunication Systems (ANTS)* (pp. 1–6).
8. Fung, C.-H. F., Tamaki, K., & Lo, H.-K. (2006). Performance of two quantum-key-distribution protocols. *Physical Review A, 73*(1), 012337.
9. Scarani, V., Bechmann-Pasquinucci, H., Cerf, N. J., Dušek, M., Lütkenhaus, N., & Peev, M. (2009). The security of practical quantum key distribution. *Reviews of Modern Physics, 81*(3), 1301.
10. Lo, H.-K., Curty, M., & Tamaki, K. (2014). Secure quantum key distribution. *Nature Photonics, 8*(8), 595–604.
11. Barbosa, G. A., Corndorf, E., Kumar, P., & Yuen, H. P. (2003). Secure communication using mesoscopic coherent states. *Physical Review Letters, 90*(22), 227901.
12. Barbosa, G. A., & van de Graaf, J. (2014). Untappable communication channels over optical fibers from quantum-optical noise. *IACR Cryptology ePrint Archive, 2014,* 146.
13. Kak, S. (2006). A three-stage quantum cryptography protocol. *Foundations of Physics Letters, 19*(3), 293–296.
14. Mandal, S., et al. (2012). Implementation of secure quantum protocol using multiple photons for communication *arXiv preprint* arXiv:1208.6198.

15. Mandal, S. (2012). *Implementation of the three-stage protocol over free space optics*. University of Oklahoma.
16. Gilbert, G., & Hamrick, M. (MITRE, 2001, June). *Constraints on eavesdropping on the BB84 protocol* (p. 6). arXiv:quant-ph/0106034v2.
17. Perkins, W. (2006, March). *Trusted certificates in quantum cryptography*. arXiv:cs/0603046v1.
18. Svozil, K. (2005). *Feasibility of the interlock protocol against man-in-the-middle attacks on quantum cryptography* (Vol. 3, No. 4, pp. 654).
19. Basuchowdhuri, P. (2006, May). *Classical authentication aided three-stage quantum protocol* (p. 7). arXiv:cs/0605083v1.
20. Thomas, J. H. (2007, June). *Variations on Kak's three stage quantum cryptography protocol* (p. 7). arXiv:0706.2888v1.
21. Punekar, N., Darunkar, B., & Verma, P. (2016, February 13). Secured optical fiber communication using polarization restoration technique and channel characterization. In *Proceedings of SPIE 9774, Next-Generation Optical Communication: Components, Sub-Systems, and Systems* V (97740F).

Chapter 7
Preliminary Security Analysis of the Multi-stage Protocol

This chapter presents a security analysis of the multi-stage protocol assessing its vulnerability to known security attacks. It shows that the multi-stage protocol can offer quantum level security under certain conditions. The material presented in this chapter is based on the authors' work previously published in [12, 13].

7.1 Introduction

A generalized multi-stage, multi-photon tolerant protocol was discussed in Chap. 6. Multi-stage protocols use arbitrary polarization states to communicate data securely between a sender and a receiver. The polarization measurement of an arbitrary and unknown polarized state results in altering the state in an irreversible way. Any such measurement produces noise in the measured state. The multi-stage protocol exploits this phenomenon to provide secure communication. This chapter assesses the vulnerability of the multi-stage protocol to a photon number splitting attack and a Trojan horse attack.

In addition, this chapter presents two approaches to calculating an upper bound on the average number of photons that can be used per pulse to exchange information while maintaining quantum-level security. The first approach is based on the assumption that an eavesdropper is faced with the problem of discriminating between two polarization states. This results in an inaccurate state estimation. The second approach is based on the assumption that a certain amount of noise will be introduced in the measurement of the state in each stage. The noise introduced may result in an incorrect value of the measured bit. The assumption we make through the second approach is that Alice and Bob are using Fock states.

Determination of an upper bound on the number of photons is important for the multi-stage protocol. It allows the multi-stage protocol to operate in the multi-photon domain while maintaining quantum-level security. This chapter shows that the average number of photons that can be used per stage of a multi-stage

P. K. Verma et al., *Multi-photon Quantum Secure Communication*,
Signals and Communication Technology,
https://doi.org/10.1007/978-981-10-8618-2_7

protocol is larger than that of its single photon counterpart. In the latter case, in BB84 and its decoy state version, an average of less than 0.5 photons per pulse is used [1–7]. A larger number of photons per pulse will result in transferring data at a higher rate and possibly over longer distances.

7.2 Background Knowledge

7.2.1 *Helstrom Discrimination*

The Helstrom formula was derived in the mid-70s to describe the minimum error probability of correct state discrimination for the case where two quantum states are used [8]. These quantum states can be either pure or mixed states, and the error probability is denoted by P_E.

This section summarizes the derivation of the probability of the correct state discrimination $P_C = 1 - P_E$ [9]. P_C is used later in this chapter to compute the number of photons that Alice and Bob can use while achieving secure communication. It is instructive to start by analyzing the two-state minimum-error measurement with the help of the method proposed in [10, 11]. Starting from,

$$P_{err} = 1 - \sum_{j=1}^{m} \eta_j \mathrm{Tr}(\rho_j \Pi_j),$$

where η_j and Π_j are the priori probabilities and detection operators respectively and ρ_j is the density operator of a given quantum system. They should satisfy the following relations.

$$\eta_1 + \eta_2 = 1 \quad \text{and} \quad \Pi_1 + \Pi_2 = I_{D_s},$$

where I is the identity operator. Then, the probability of getting an erroneous result in the measurement is given by [9].

$$P_{err} = 1 - \sum_{j=1}^{2} \eta_j \mathrm{Tr}(\rho_j \Pi_j) = \eta_1 \mathrm{Tr}(\rho_1 \Pi_2) + \eta_2 \mathrm{Tr}(\rho_2 \Pi_1).$$

This can be alternatively expressed as

$$P_{err} = \eta_1 + \mathrm{Tr}(\Lambda \Pi_1) = \eta_2 - \mathrm{Tr}(\Lambda \Pi_2).$$

where we introduced the Hermitian operator,

$$\Lambda = \eta_1 \rho_1 + \eta_2 \rho_2 = \sum_{k=1}^{D_s} \lambda_k |\phi_k\rangle\langle\phi_k|.$$

Here the states $|\phi_k\rangle$ denote the orthonormal eigenstates corresponding to the eigenvalues λ_k of the operator Λ. By using the spectral decomposition of Λ, we get the representations:

$$P_{err} = \eta_1 + \sum_{k=1}^{D_s} \lambda_k \langle \phi | \Pi_1 | \phi_k \rangle = \eta_2 + \sum_{k=1}^{D_s} \lambda_k \langle \phi_k | \Pi_2 | \phi_k \rangle. \tag{7.1}$$

Now the optimization task consists of determining the specific operators Π_1, or Π_2, respectively, that minimize the right-hand side of Eq. (7.1) under the constraint that,

$$0 \leq \langle \phi_k | \Pi_j | \phi_k \rangle \leq 1, \quad (j = 1, 2)$$

for all eigenstates $|\phi_k\rangle$. The latter requirement is needed because $\mathrm{Tr}(\rho \Pi_j)$ denotes a probability for any ρ. From this constraint and from Eq. (7.1), it follows that the smallest possible error probability, $P_{err}^{min} = P_E$, can be achieved when the detection operators are selected in such a way $\langle \phi_k | \Pi_1 | \phi_k \rangle = 1$ and $\langle \phi_k | \Pi_2 | \phi_k \rangle = 0$ are fulfilled for eigenstates belonging to negative eigenvalues. Meanwhile eigenstates corresponding to positive eigenvalues obey the equations $\langle \phi_k | \Pi_1 | \phi_k \rangle = 0$ and $\langle \phi_k | \Pi_2 | \phi_k \rangle = 1$. Thus, the optimum detection operators can be written as,

$$\Pi_1 = \sum_{k=1}^{k_0 - 1} |\phi_k\rangle\langle\phi_k| \quad \Pi_1 = \sum_{k=k_0}^{D_s} |\phi_k\rangle\langle\phi_k|$$

By inserting the optimum detection operators into the minimum error probability is found to be [11].

$$P_E = \eta_1 - \sum_{k=1}^{k_0 - 1} |\lambda_k| = \eta_2 - \sum_{k_0}^{D_s} |\lambda_k| \tag{7.2}$$

Taking the sum of these two alternative representations and using $\eta_1 + \eta_2 = 1$, P_E is represented by,

$$P_E = \frac{1}{2}(1 - \sum_k |\lambda_k|) = \frac{1}{2}(1 - \mathrm{Tr}(|\Lambda|)$$

where $|\Lambda| = \sqrt{\Lambda^\dagger \Lambda}$. Together with (7.1) this yields the well-known Helstrom formula [8] for the minimum error probability in discriminating ρ_1 and ρ_2,

$$P_E = \frac{1}{2}(1 - \mathrm{Tr}|\eta_2 \rho_2 - \eta_1 \rho_1|)$$

In the special case that the states to be distinguished are the pure states $|\phi_1\rangle$ and $\langle\phi_2|$, this expression reduces to [8].

$$P_E = \frac{1}{2}\left(1 - \sqrt{1 - 4\eta_1\eta_2|\langle\phi_1|\phi_2\rangle|^2}\right)$$

Giving rise to a probability of correct measurement given by,

$$P_C = \frac{1}{2}\left(1 + \sqrt{1 - 4\eta_1\eta_2|\langle\phi_1|\phi_2\rangle|^2}\right) \tag{7.3}$$

Equation (7.3) is used in Sect. 7.2.1 to derive the number of photons needed if Eve was to carry a photon-number-splitting (PNS) attack on the three-stage protocol. In this case we consider the capability of Eve to distinguish between two quantum states. It is worth noting that Sect. 7.3.1 uses the condition $\eta_1 = \eta_2 = \frac{1}{2}$ that denotes the optimum discrimination strategy [9].

It is important to note that the calculations used Helstrom discrimination assume Eve is measuring the state of photons sent on each leg of the protocol seperatley. However, in the rest of this book we take a different approach to calculate the average number of photons to be used during the communication. This approach consider the three stages of the protocol together. Hence the difference in the average calculated in the section and the rest of the book can be noted.

7.3 Photon Number Splitting Attack (PNS)

A photon number splitting attack in the context of a multi-stage multi-photon protocol requires an eavesdropper to siphon off a certain number of photons at each stage of the protocol in order to be able to recover the message sent over that stage. In the case of the three-stage protocol, Eve would siphon off n photons at each of the three legs. The PNS attack in the case of a three-stage protocol is shown in Fig. 7.1. Alice transfers the pulse after changing the polarization of the input to $X + \alpha$, where X represents the polarization angle depending on whether the input bit was a ')' or a '1', and alpha is the polarization angle of her choice. Eve steals a few photons from the pulse as shown in Fig. 7.1 so that she could measure the polarization of the photons stolen. The process is repeated at each stage of the protocol. Measurement of the polarization at each stage is essential since the information from only one stage will not be sufficient to get the original encoded polarization angle, X.

It is well known that as the number of photons in the optical beam increases, Eve's ability to accurately measure the polarization of the beam increases, and vice versa. In particular, this is due to the fact that the fidelity with which Eve can measure the state of polarization increases with the number of photons split from the beam.

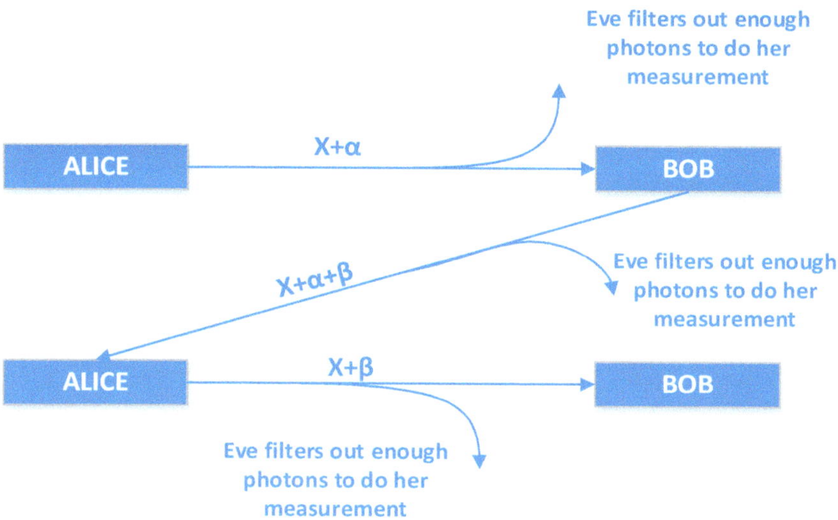

Fig. 7.1 Photon number splitting attack on the three-stage protocol

Assume that Eve successfully siphoned an equal number of photons off each of the three stages of the protocol. Knowing that a measurement of the value of the polarization angle would introduce noise, Eve will not be able to deduce the exact value of the incoming polarization state from the measured states of polarization. If the measurement on each stage introduces an equal amount of uncertainty to the actual angle of polarization sent over the channel, Eve will have in her possession, in the worst case scenario, $\phi + 3\Delta\phi$, where $\Delta\phi$ is the uncertainty added by the measurement on one stage. This logic can be applied to both the cases of state discrimination as well as Fock state measurement.

Eve will have no precise information about ϕ when [12]:

$$3\langle\Delta\phi\rangle \geq \frac{\pi}{4}, \quad \text{or equivalently} \quad \langle\Delta\phi\rangle \geq \frac{\pi}{12}.$$

7.3.1 Helstrom Discrimination

As the number of photons in a beam increases, the level of vulnerability of the protocol towards a PNS attack increases. In this section, we use the probability P_C calculated in Sect. 7.2.1 in order to calculate the minimum average number of photons Eve needs to siphon off for her measurement. This number of photons per stage is the maximum number per stage that should be used in the three-stage protocol to provide secure communication. It is calculated in terms of the probability P_C, in such a way that at each stage $\langle\Delta\phi\rangle \geq \frac{\pi}{12}$ is introduced giving rise to a

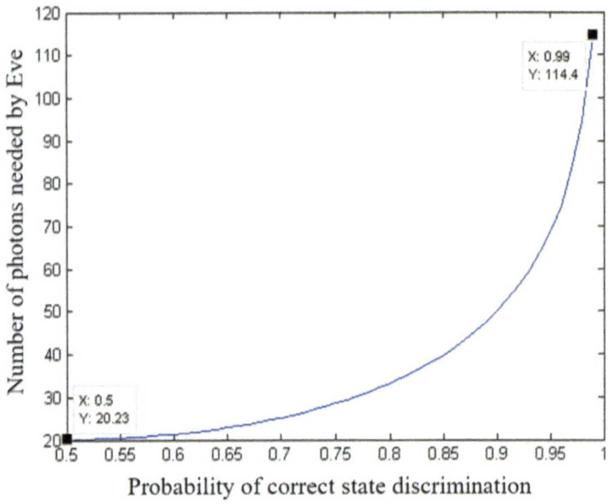

Fig. 7.2 Interplay between the number of photons and P_C

total difference of more than or equal to $\frac{\pi}{4}$ between the original angle sent over the channel and the angle Eve has by the end of the protocol execution.

Figure 7.2 illustrates the interplay between the number of photons per stage in the beam and P_C. The three-stage protocol is unconditionally secure as long the number of photons per stage used is less that depicted by the blue curve in Fig. 7.2. We take y as the actual number of photons per stage sent over the channel.

The interplay between the number of photons and M that can be derived using P_C. Let's call M the number of achievable polarization states at angle $\frac{k\pi}{M}$ with $k \in [0 \ldots N-1]$. The possible states of y photons can be written as follows [13].

$$|\phi_k\rangle = |k, y\rangle = \left(\frac{|0\rangle + e^{2ik\pi/M}|1\rangle}{\sqrt{2}}\right)^{\otimes y}. \tag{7.4}$$

The probability of confusing two different angles k and k' is the scalar product given by:

$$\langle k', y|k, y\rangle = \left[\cos\left(\frac{(k-k')\pi}{M}\right)\right]^y. \tag{7.5}$$

If the problem is restricted to discrimination between two angles k and $k+1$, the probability of success of a Helstrom discrimination is given by [14].

$$P_C = \frac{1}{2}\left(1 + \sqrt{1 - |\langle k', y|k, y\rangle|^2}\right) \tag{7.6}$$

$P_c \geq 1 - \epsilon$ where $\epsilon \ll 1$, thus

$$\left(\cos\frac{\pi}{M}\right)^{y} \leq 2\sqrt{\epsilon(1-\epsilon)},$$

$$y \geq \frac{\ln\epsilon + 2\ln 2 + \ln(1-\epsilon)}{2\ln\left(\cos\frac{\pi}{M}\right)}. \tag{7.7}$$

When $M \gg 1$ we have $\cos\frac{\pi}{M} \cong 1 - \frac{1}{2}\left(\frac{\pi}{M}\right)^2$ and therefore,

$$\log\left(\cos\frac{\pi}{M}\right) \cong -\frac{1}{2}\left(\frac{\pi}{M}\right)^2,$$

$$y \geq M^2\frac{2}{\pi^2}(-\ln\epsilon - \ln 2 - \ln(1-\epsilon)).$$

In other words, this implies that the number of photons needed to discriminate between two polarization states is of the order M^2.

The calculations above are meant to distinguish between two states differing by an angle of $\frac{\pi}{M}$. In order to be applicable in the case of the three-stage protocol, the uncertainty introduced by the state discrimination used should be at least equal to $\frac{\pi}{12}$. Setting $M = 12$ we get the following relation.

$$y \geq 12^2\frac{2}{\pi^2}(-\ln\epsilon - \ln 2 - \ln(1-\epsilon)).$$

Once again y in this equation is the minimum number of photons needed by Eve and in the meanwhile it is the maximum number of photons that should be sent over the channel. This in turn means,

$$y < 12^2\frac{2}{\pi^2}(-\ln\epsilon - \ln 2 - \ln(1-\epsilon)) \tag{7.8}$$

7.3.2 Fock States

In this section, the following calculations are used to find a generalized relation between $\langle\Delta\phi\rangle$, the average angular deviation that a measurement done at Eve's side has, and the minimum number of photons N Eve needs for her measurement. We consider the case of Fock states [15] or number states discussed in Chap. 2 of this book to find the variance induced while measuring the polarization state using H-mode and V-mode. H-mode and V-mode represent measurements with respect to a horizontal or a vertical reference, respectively.

In our analysis, we represent our states as number states or Fock states. We assume that the vacuum state $|\psi_0\rangle = |0\rangle$ is already normalized. Then the Nth Fock state is given by,

$$|\psi_N\rangle = c_N (b^\dagger)^N |0\rangle, \tag{7.9}$$

where $b^\dagger = \cos\phi a_H^\dagger + \sin\phi a_V^\dagger$ with a_H and a_V being the field operators for horizontal and vertical polarizations respectively, and the normalization constant $c_N = 1/\sqrt{N!}$. The binomial expansion of $|\psi_N\rangle$ is given by,

$$|\psi_N\rangle = \sum_{k=0}^{N} \sqrt{\frac{N!}{k!(N-k)!}} \cos^{\mathrm{k}}\phi \sin^{(N-k)}\phi |k\rangle_H |N-k\rangle_V \tag{7.10}$$

Since our decoding scheme is based on intensity detection, we derive Malus' Law for our analysis. The average photon number in H-mode is given by the mean intensity distribution:

$$\langle I_H\rangle = \sum_{k=0}^{\infty} k|\langle k|\psi_N\rangle_H|^2 = \sum_{k=0}^{N} k\left[\frac{N!}{k!(N-k)!}\right]\left(\cos^2\phi\right)^k\left(\sin^2\phi\right)^{N-k} = N\cos^2\phi \tag{7.11}$$

Similarly, $\langle I_V\rangle = N\sin^2\phi$. Then we find that the normalized intensities become,

$$\frac{I_H}{I_H + I_V} = \cos^2\phi, \tag{7.12}$$

and

$$\frac{I_V}{I_H + I_V} = \sin^2\phi. \tag{7.13}$$

The variance in the intensity measurement in H-mode done by Eve can be found as follows:

$$\langle I_H^2\rangle - \langle I_H\rangle^2 = \frac{N}{4}\sin^2 2\phi. \tag{7.14}$$

Thus the normalized deviation of the H-mode is given by:

$$\frac{\sqrt{\langle I_H^2\rangle - \langle I_H\rangle^2}}{I_H + I_V} = \frac{|\sin 2\phi|}{2\sqrt{N}} \tag{7.15}$$

It can be shown that the normalized deviation of the V-mode is identical to that of the H-mode.

In order to find $\Delta\phi$ we take derivative of both sides of Eq. (7.12) as follows:

$$\Delta\left(\cos^2\phi\right) = \Delta\left(\frac{I_H}{I_H + I_V}\right) = \frac{\Delta I_H}{N},$$

since $I_H + I_V = N$ for Fock states. This gives,

$$\Delta\phi = \frac{1}{2\sqrt{N}}.$$

Thus, to have $\langle\Delta\phi\rangle \geq \frac{\pi}{12}$, the number of photons $N \leq 1.91$. Therefore, the total number of photons in the beam should be $3N \leq 5.73$.

7.4 Trojan Horse Attack

A Trojan horse attack [16] that can be launched by Eve is shown in Fig. 7.3. During this attack, Eve injects an optical beam with a known state of polarization right before Bob applies his transformation U_B. Then Eve filters out her photons at the second stage of the protocol. By measuring and comparing the polarization states of the injected and the recovered photons, Eve can now estimate the polarization transformation introduced by Bob (U_B). During the execution of the protocol in the third stage, Eve can launch an intercept and resend attack. Having a prior knowledge of the transformation U_B, she will be able to recover the value of X being sent.

To defeat a Trojan horse attack, Alice and Bob should actively monitor the intensity of the optical beam at each end. This can be implemented using auxiliary detectors and active monitoring of any incoming beam. Furthermore, one can use an "optical fuse" to monitor any increase in the beam intensity over the channel. An increase in the beam intensity takes place in case Eve injects an optical beam with a slightly higher intensity than the one she siphoned off. It is worth noting that the intensity threshold at which the protocol is operating is preset before the onset of the implementation and is mainly determined by the number of photons used by

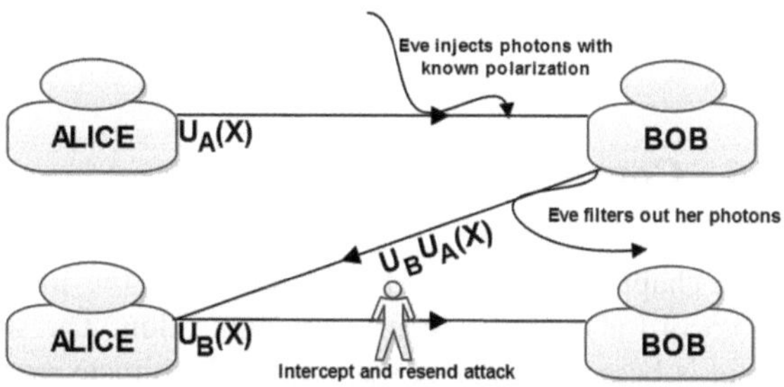

Fig. 7.3 Diagram of a Trojan horse attack on the three-stage protocol

Alice and Bob. Furthermore, the noise on the channel can be characterized prior to any data exchange; thus, Alice and Bob can know whether the intensity fluctuations were due to eavesdropping or to noise.

7.5 Hardware Countermeasures

In order to limit the vulnerability of a system implementing the three-stage protocol to the attacks discussed above one can implement the following hardware countermeasures:

(1) Limit the maximum number of photons to a value less than that proposed in Sect. 7.3.2. We call this method the multi-photon tolerant approach. Using this approach, one can guarantee that an eavesdropper cannot measure the actual polarization state over a given stage of the protocol; thus, the message X is transmitted at a secure level while using multiple photons in each pulse. The use of a single photon to encode each bit maintains the security level of the BB84 protocol while removing its major limitations. Therefore, loosening the limits on the distance, data rates, and single photon generation and detection. It is worth noting that we use the value $3N \leq 5.73$ found in Sect. 7.3.2; this is due to the assumption that Eve will use the method where she needs the least number of photons in order to launch her attack.

(2) Actively monitor the intensity of the optical beam at each receiving end. This can be implemented using auxiliary detectors and active monitoring of any incoming light. Furthermore, one can use an optical fuse to monitor any increase in the beam intensity over the channel. An increase in the beam intensity takes place in case Eve injects an optical beam with slightly higher intensity than the one she siphoned.

(3) A "door" function can be applied as well; this is an effective countermeasure against a Trojan horse attack. This door can be thought of as a narrow band filter that will disallow Eve from injecting light at a wavelength different than that used during communication. Therefore, Eve's presence can be detected through intensity measurements that can be carried out as proposed in 2 above. Furthermore, the door function can open access to the channel only once a legitimate signal has been sent through the channel. In other words, the system will be in an idle state as long as Alice and Bob are not sending any information over the channel.

7.6 Conclusion

This chapter has analyzed the security of the multi-stage, multi-photon tolerant protocol for quantum secure communication. The security of the multi-stage protocol is based on the fact that while a legitimate receiver only needs to distinguish between two orthogonal polarization states, an intruder has to distinguish among an indefinite number of possible polarization states. In other words, while the intended recipient would need only one photon to do so, an intruder would need to siphon off a minimum number N of photons per stage. This chapter has discussed the calculations needed to determine the number of photons N using two different approaches. The first is using Helstrom discrimination while the second is using the assumption that Alice and Bob make use of Fock states to implement the three-stage protocol. It is worth noting that the protocol implementation must use no more than the minimum theoretically secure number of photons. The results presented in this chapter set an upper bound on the number of photons at $3N \leq 5.73$ in the case of the three-stage protocol. As long as the average number of photons that Eve can siphon off from each leg is less than N, the communication between the parties is expected to be secure from a photon number splitting attack.

The chapter has also assessed the security of the generalized multi-stage protocol to the Trojan horse attack and the additional measures for defying such attacks have been proposed. The proposed security analysis is discussed in the case of the three-stage multi-photon protocol; however, it should be noted that this analysis can be generalized to any case of the multi-stage protocol. Furthermore, a comparison between the BB84 protocol and the multi-stage protocol has been presented. The proposed security analysis discussed demonstrates that a multi-photon protocol can provide secure communication while using a larger average number of photons. This in turn enhances the performance of the multi-stage protocol compared to its single-photon counterparts.

References

1. Lo, H.-K., & Chau, H. F. (1999). Unconditional security of quantum key distribution over arbitrarily long distances. *Science, 283*(5410), 2050–2056.
2. Shor, P. W., & Preskill, J. (2000). Simple proof of security of the BB84 quantum key distribution protocol. *Physical Review Letters, 85*(2), 441.
3. Gottesman, D. et al. (2004) Security of quantum key distribution with imperfect devices. In *Information Theory, 2004. ISIT 2004. Proceedings. International Symposium on*, IEEE.
4. Kraus, B., Branciard, C., & Renner, R. (2007). Security of quantum-key-distribution protocols using two-way classical communication or weak coherent pulses. *Physical Review A, 75*(1), 012316.
5. Lo, H.-K., Ma, X., & Chen, K. (2005). Decoy state quantum key distribution. *Physical Review Letters, 94*(23), 230504.
6. Ma, X., et al. (2005). Practical decoy state for quantum key distribution. *Physical Review A, 72*(1), 012326.
7. Peng, C.-Z., et al. (2007). Experimental long-distance decoy-state quantum key distribution based on polarization encoding. *Physical Review Letters, 98*(1), 010505.
8. Helstrom, C. W. (1976). *Quantum detection and estimation theory*. Cambridge: Academic press.
9. Bergou, J. A. (2010). Discrimination of quantum states. *Journal of Modern Optics, 57*(3), 160–180.

10. Fuchs, C. A. (1996) *Distinguishability and accessible information in quantum theory*. arXiv preprint quant-ph/9601020.
11. Herzog, U. (2004). Minimum-error discrimination between a pure and a mixed two-qubit state. *Journal of Optics B: Quantum and Semiclassical Optics, 6*(3), S24.
12. El Rifai, M., Chan, K. W. C., Verma, P. K. (2015) *Multi stage quantum secure communication using polarization hopping*. Security and Communication Networks.
13. El Rifai, M., Verma, P. K. (2015) *Quantum secure communication using a multi-photon tolerant protocol*. In *SPIE OPTO, International Society for Optics and Photonics*.
14. Grosshans, F. (2012) *How many photons does it take to measure a linear polarization?* Available November 4, 2015 from: http://physics.stackexchange.com/questions/22575/how-many-photons-does-it-take-to-measure-a-linear-polarization.
15. Fox, M. (2006). *Quantum optics: An introduction* (Vol. 6). Oxford: Oxford University Press.
16. Gisin, N., et al. (2006). Trojan-horse attacks on quantum-key-distribution systems. *Physical Review A, 73*(2), 022320.

Chapter 8
Security Analysis of the Multi-stage Protocol

This chapter analyzes intercept-and-resend and photon number splitting attacks in the multi-stage multi-photon protocol. It lays down the conditions under which the multi-stage multi-photon protocol can approach the strength of a quantum-secure protocol. The material presented in this chapter is based on the authors' work previously published in [16].

8.1 Introduction

The previous chapters have presented a multi-stage, multi-photon protocol based on the double-lock cryptography. As mentioned, the multi-stage protocol exploits the asymmetry between the detection strategies of the legitimate users and the eavesdropper. In this chapter, we study the security of the multi-photon protocol using coherent states by demonstrating its security against the intercept-resend (IR) attack, the photon-number-splitting (PNS) attack, and the man-in-the-middle (MIM) attack. It is found that the mean photon number of the coherent pulses can generally be greater than 1. The protocol thus has the potential to offer quantum security using detectors of less stringent specification.

Security in the multi-photon, multi-stage protocol is due to the asymmetry in the detection strategies of the intended receiver and the eavesdropper. This asymmetry is provided by the advantage creation akin to that utilized in the optimal quantum receiver in the Y00 (or $\alpha\eta$) protocol [1] and the keyed communication in quantum noise (KCQ) method [2].

This chapter computes the secure key transfer rate in terms of the optimal error probability of eavesdropping. In addition, it devises an authentication method to check for a potential man-in-the-middle attack that could be launched by an eavesdropper. Furthermore, this chapter estimates the modified secure key rate achievable in the presence of such an attack. It is worth noting that in case the key/message

P. K. Verma et al., *Multi-photon Quantum Secure Communication*,
Signals and Communication Technology,
https://doi.org/10.1007/978-981-10-8618-2_8

expansion protocol proposed in Chap. 5 is used, the three-stage protocol acts as a quantum communication protocol rather than key distribution protocol. However, in the authentication method developed in this chapter, the three-stage protocol is used as a key distribution protocol. In addition, an estimate of the maximal mean photon number when the channel loss is taken into account is calculated. This chapter also provides an analysis of the security of the simplest form of the multi-stage protocol, the three-stage protocol. The relationship between the error probabilities and the mean number of photons in the channel is developed.

The amount of polarization rotations that both sides (Alice and Bob) select for the information bits is an arbitrary and independent value which varies from 0 to π degrees. It should be noted that practically one must also take into account the difficulties with maintaining fidelity in the presence of noise. In the following analysis, such a requirement is ignored and it is assumed that Alice and Bob can maintain perfect alignment in their bases for simplicity.

8.2 Intercept-Resend (IR) and Photon Number Splitting (PNS) Attacks

First, consider the situation that the communication between Alice and Bob is “authenticated”, i.e., Alice knows that the information she sends out passes through Bob in the intermediate step, and vice versa. Under this assumption, Eve can launch intercept-resend (IR) attacks, or more importantly, the photon-number-splitting (PNS) attack.

The main difference between IR and PNS is that, in IR all the photons are being taken out by Eve and she then resends any photon state to Bob. On the other hand, under the PNS attack, the number of photons Bob receives is less than that in the original pulse. Such a loss of photons practically could be due to the channel loss but, in this analysis, it is attributed to the action of the intruder.

If we further restrict ourselves to the situation of incoherent attacks, Eve is required to perform measurements before the classical post processing. Therefore, under the IR attack, the polarization states of the pulses resent by Eve usually are different from those she intercepts because of the measurement process, with the difference depending on the number of photons she receives. The security against the IR attack can be estimated by assuming that the polarization states of the resent pulses are the same as those before the pulses are intercepted by Eve. This is an overestimation of the ability of Eve. Nevertheless, it enables us to analyze IR and PNS attacks using the same formalism. In addition, Eve’s information I_{EA} and I_{EB} become identical, where $I_{EA} = \max_{\text{Eve}} I(E:A)$ is the maximal mutual information between Eve and Alice with a similar expression for $I(E:B)$.

For the IR and PNS attacks on the three-stage protocol, Eve only needs to measure the polarization angles of any two stages. Then she can extract the bit value by orienting her measurement device in the third stage according the angles of the first and second stages.

More specifically, suppose the polarization angles of the three stages of the protocol are denoted by $\phi_1 = X + \alpha$, $\phi_2 = X + \alpha + \beta$ and $\phi_3 = X + \beta$, where X is the information bit angle (0 or $\frac{\pi}{2}$), and α and β are the angles associated with Alice's and Bob's unitary transformations. Then the corresponding angles estimated by Eve for the first two stages are written as $\hat{\phi}_1 = \hat{X} + \hat{\alpha}$ and $\hat{\phi}_2 = \hat{X} + \hat{\alpha} + \hat{\beta}$. As a result, $\hat{\beta} = \hat{\phi}_2 - \hat{\phi}_1$. In order for Eve to obtain useful information, she requires that the error in determining β should not be too large. Since X is a binary random number of either 0 or $\frac{\pi}{2}$, Eve will determine the bit value erroneously if $\left|\hat{\beta} - \beta\right| = \left|(\hat{\phi}_2 - \hat{\phi}_1) - (\phi_2 - \phi_1)\right| > \pi/4$. The error probability of Eve is then given by

$$P_e(N_1, N_2) = \int\limits_S d\hat{\phi}_1\, d\hat{\phi}_2\, d\phi_1 d\phi_2\; P(\phi_1)P_1(\hat{\phi}_1|\phi_1, N_1)P(\phi_2)P_2(\hat{\phi}_2|\phi_2, N_2) \quad (8.1)$$

where $P(\phi) = 1/2\pi$, is the prior distribution of Alice's (Bob's) rotation angle and $P_i(\hat{\phi}_i|\phi_i, N_i)$ is the conditional probability of determining $\hat{\phi}_i$ given the angle ϕ_i and the mean photon number N_i that is accessible by Eve. The integration domain S corresponds to the region where the condition $\left|(\hat{\phi}_2 - \hat{\phi}_1) - (\phi_2 - \phi_1)\right| > \pi/4$ is satisfied. The mutual information $I(E:A)$ is given by $I(E:A) = 1 - h(P_e)$.

Consider a three-stage protocol using coherent states of mean photon number N. First of all, Alice should randomize the phases of the coherent states to avoid Eve exploiting the phase information [3]. In this case, the quantum state is described by a density matrix with photon number following the Poisson distribution with parameter N.

To obtain a bound of the secure key rate, one has to estimate Eve's maximal information. This involves an optimal measurement strategy to obtain the conditional probability $P_i(\hat{\phi}_i|\phi_i, N_i)$. Bagan et al. [4] has given a detailed comparison of the estimation of the polarization state of a finite number of photons using the collective and local measurements. Instead of an optimal polarization measurement, in the following we consider a simple strategy that Eve performs polarization analysis with a fixed basis, denoted as horizontal and vertical, that is the same as Alice and Bob's basis. Such a fixed basis measurement is generally not optimal. Nevertheless, we additionally assume that Eve can determine the polarization angle correctly using a single basis only, instead of two bases that are required for the polarization states on a circle of the Poincaré sphere. This is accomplished by attributing Eve's measured polarization in the correct quadrant as the original polarization in the numerical calculations below. This procedure effectively doubles the number of photons available to Eve for the estimation, and the fidelity obtained is generally even better than that using optimal collective measurements

With the measurement strategy mentioned above, the probability distributions of Eve's numbers of horizontal and vertical photons in the three stages are given by

$$P_i(n_{H,i}, n_{V,i}|\phi_i, N_i) = \frac{e^{-N_i}}{1 - e^{-N_i}} \frac{(N_i \cos^2 \phi_i)^{n_{H,i}} (N_i \sin^2 \phi_i)^{n_{V,i}}}{n_{H,i}! n_{V,i}!} \tag{8.2}$$

for $i = 1, 2, 3$, where N_i is the mean number of photons in stage i that is accessible by Eve. Here the continuous variable $\hat{\phi}_i$ in Eq. (8.1) is replaced by the discrete variables $n_{H,i}$ and $n_{V,i}$. Then ϕ_i can be estimated from the numbers of photons detected in the vertical port $(n_{V,i})$ and the horizontal port $(n_{H,i})$ of the polarization analyzer by $\tan^2 \hat{\phi}_i = n_{V,i}/n_{H,i}$. Note that in Eq. (8.2), $n_{H,i}$ and $n_{V,i}$ cannot be zero simultaneously, for this gives no information to Eve about the angle ϕ_i. Also we assume N_i is known to Eve.

For the PNS attack, Eve's best strategy without causing errors to Bob's received bits will be to take $N_1 = N_2 \sim \frac{N}{2}$ if Bob did not monitor the photon statistics. Nevertheless, we require that Bob monitors the number of incoming photons so that Eve cannot probe Alice and his devices with very bright pulses. For the IR attack, we can consider $N_1 = N_2 \sim N$. This corresponds to the optimal situation for Eve when the channel is assumed to be lossless. For a lossy channel with transmittance t, we consider $N_1 = N$ and $N_2 = tN$ for IR and $N_1 = (1-t)N$ and $N_2 = (1-t)tN$ for PNS. Figure 8.1 shows the case of a IR attack versus a PNS attack. Figure 8.2 gives the plots of P_e as a function of the mean photon number N. It is seen in Fig. 8.2 that even at the mean photon number $N = 10$, there is considerable error in Eve's estimated values of the true bit values.

As mentioned previously, Alice and Bob need to monitor the number of incoming photons to deny Eve from injecting a very bright beam to probe their encoding devices. The presence of Eve is revealed if Alice and Bob also check the photon number distribution and detect any loss or change of the distribution. Eve could compensate the photon loss in the channel by injecting photons of arbitrary polarizations or at the angles $\hat{\phi}_i$, as in the IR attack. Nevertheless, this introduces extra error in her determination of $\hat{X}$ as well as error in Bob's bits. In addition, the IR attack in fact induces errors to the bit values obtained by Bob. The estimation of the rotation angle error is addressed by the authentication process which specifically handles the man-in-the-middle attack in the next section.

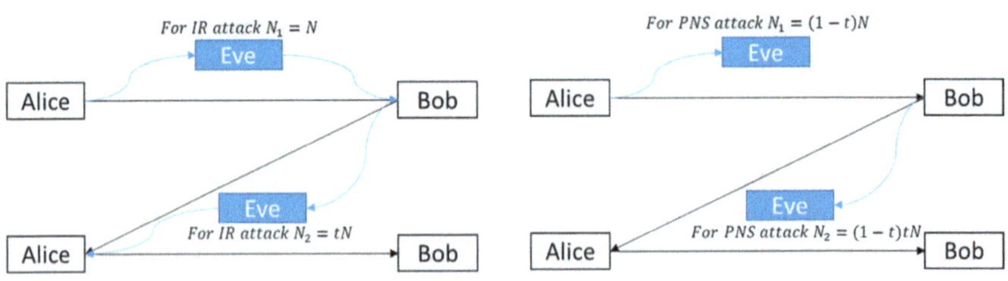

Fig. 8.1 IR versus PNS attack on a three-stage protocol

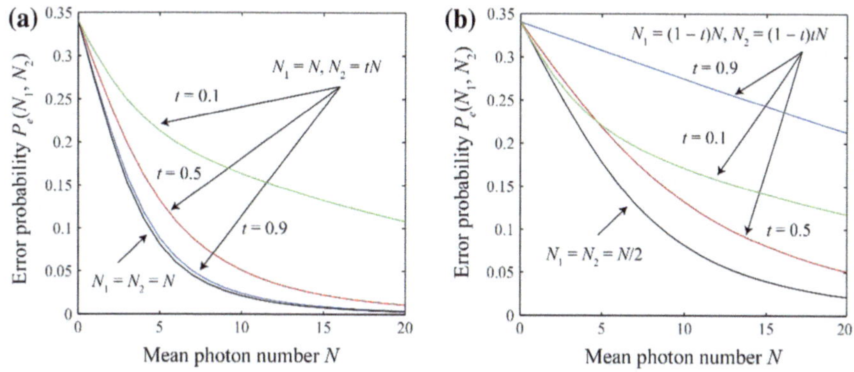

Fig. 8.2 Plots of the **a** IR and **b** PNS error probabilities of Eve as functions of the mean number of photons N

8.3 Authentication

The three-stage protocol can be compromised entirely if Eve launches the man-in-middle (MIM) attack as depicted in Fig. 8.3. Here Eve impersonates Bob to extract the true bit value perfectly. She also impersonates Alice to send the bit angle X together with the unperturbed angle β back to Bob, so that Bob receives the bit without error and hence cannot catch Eve. In such an MIM attack, Eve totally separates the quantum communication between Alice and Bob. Therefore the attack could be revealed if authentications are made by Alice and Bob to guarantee the locks are legitimate, i.e., they are the true users who applied the rotation angles on the pulses they received in the three stages.

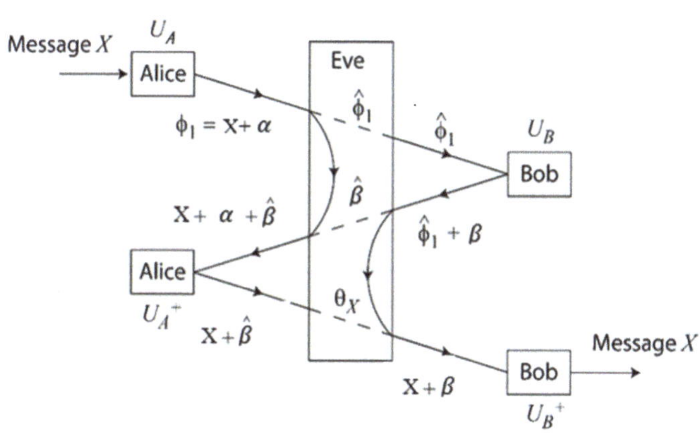

Fig. 8.3 Schematic diagram of the three-stage protocol under the man-in-the-middle (MIM) attack

In principle, authentication can be performed perfectly if Alice and Bob could retain the photons in steps 2 and 3 of the protocol above until the end of the key exchange, which can be accomplished by using quantum memories [5, 6] or slow light technologies [7]. More practically they need to perform measurements to determine the parameters of the transformations during the key exchange. At the end of the key exchange, they check their measured values against the true values (step 6). It should be noted that we assume Alice and Bob are authenticated for exchanging classical information on a public channel. This will rule out the chance that Eve is also in the middle when Alice and Bob try to compare the measurements.

We consider that the transmittance of the quantum channel is t. If Alice sends pulses with a mean photon number of N, Bob expects to receive pulses with mean photon number tN in the first stage and t^3N in the third stage, and Alice expects to receive pulses with mean photon number t^2N in the second stage. Therefore, for the MIM attack, Eve can extract a mean photon number of $(1-t^2)N$ to obtain the estimate $\hat{\phi}_1$ and a mean photon number of $t(1-t^2)N$ to obtain the estimate $\hat{\beta}$. Eve then uses these two angles to impersonate Alice and Bob simultaneously. If Bob uses the pulse for authentication instead of the normal three-stage, the angle $\tilde{\phi}_1$ he measures conditioned on ϕ_1 will have a distribution given by $P(\tilde{\phi}_1|\hat{\phi}_1, tN)$ $P(\hat{\phi}_1|\phi_1,(1-t^2)N)$. Here ϕ_1 is announced to Bob by Alice at the end of the protocol. Using this angle, Bob can guess X with an error probability of

$$P_e^{\text{Auth,MIM}}(t,N) = \int\limits_{|\phi_1-\tilde{\phi}_1|>\pi/4} d\phi_1 d\hat{\phi}_1 d\tilde{\phi}_1 \; P(\tilde{\phi}_1|\hat{\phi}_1,tN)P(\hat{\phi}_1|\phi_1,(1-t^2)N)\; P(\phi_1) \quad (8.3)$$

On the other hand, in the normal operation when the MIM attack is not present, Bob's error probability is instead given by

$$P_e^{\text{Auth,normal}}(t,N) = \int\limits_{|\phi_1-\tilde{\phi}_1|>\pi/4} d\phi_1 d\tilde{\phi}_1 \; P(\tilde{\phi}_1|\phi_1,tN)P(\phi_1) \quad (8.4)$$

It is remarked that Eqs. (8.3) and (8.4) manifest the fact that, like Eve, Bob and Alice cannot estimate the polarization angles with certainty in the middle of the three-stage protocol because the photons are not in orthogonal states.

Numerical simulations were performed using the measurement scheme described in the last section. Figure 8.4 shows the two error probabilities as functions of the mean photon number N for different values of the transmittance t. In addition, $P_e^{\text{Auth,normal}}(t,N)$ is found analytically to be

$$P_e^{\text{Auth,normal}}(t,N) \approx \frac{2}{\pi(1-e^{-N})}\left[\int_{\frac{\pi}{4}}^{\frac{\pi}{2}} \sum_{n_H=1}^{\infty} P_1(n_H,0|\phi_1,tN)d\phi_1 + \int_{0}^{\frac{\pi}{4}} \sum_{n_V=1}^{\infty} P_1(0,n_V|\phi_1,tN)d\phi_1\right]$$
$$= \frac{4}{\pi}\int_0^{\frac{\pi}{4}} \frac{1-e^{tN\sin^2\phi_1}}{1-e^{tN}}d\phi_1 = \frac{e^{\frac{tN}{2}}[I_0(\frac{tN}{2}) - L_0(\frac{tN}{2})]-1}{e^{tN}-1} \tag{8.5}$$

where In(x) is the modified Bessel function of the first kind and Ln(x) is the modified Struve function.

It is noted in Fig. 8.4 that at small N the error probabilities tend to the constant values $P_e^{\text{Auth, norm}} \to 2^{-1} - \pi^{-1}$ and $P_e^{\text{Auth, MIM}} \to 0.5$ whereas both probabilities tend to zero at large N. When the transmittance decreases, the two error probabilities converge to each other at a smaller N. In addition, the difference $P_e^{\text{Auth, MIM}} - P_e^{\text{Auth, norm}}$ approaches an asymptotic form when $t \to 0$, which is non-negligibly greater than zero for $N < 4$.

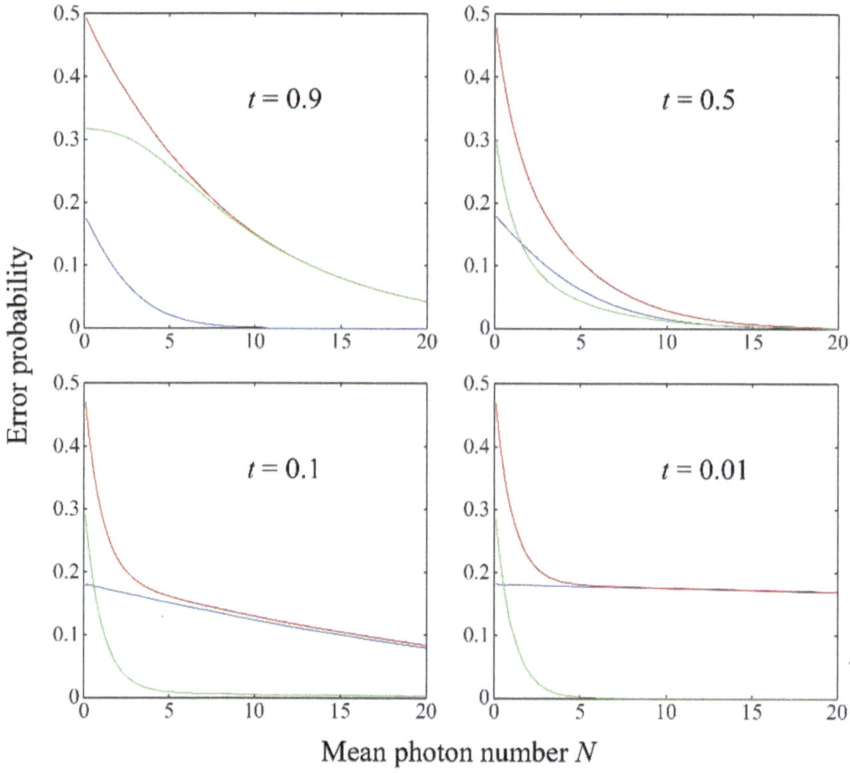

Fig. 8.4 Bob's error probabilities in the estimation of X for the normal three-stage operation (blue lines) and under the MIM attack (red lines) at different values of the channel transmittance t. The green lines denote the differences between the two error probabilities

8.4 Amplification Attack

So far we have only focused on the situation where Eve makes direct measurement using the photons that she siphons off from the quantum channel. Generally she can do more with her photons. An important class of attacks is the one that Eve amplifies the quantum states that she extracts from the channel. Figure 8.5 shows the case of an amplification attack on the three-stage protocol. This kind of attack is linked to the foundation of the three-stage protocol, that is whether she can find out the angles α and β, which are open to her eavesdropping, with high precision. In fact, the purpose of using finite number of photons in the channel is to limit Eve's precision of measurement.

It is well known that the amplification of a quantum state must also accompany with the amplification of the noise [8]. For the implementation with coherent states discussed in this chapter, Eve does not gain anything by amplifying the signal. Even with the use of squeezed states, Fock states or entangled states to resend pulses to Bob and Alice, the intensity check by Alice and Bob will introduce vacuum noise to Eve's probes, and Eve's information gain may only be modest.

On the other hand, it has recently been shown that noiseless amplification of a quantum state is possible if a perfect guarantee of success is not required, unlike the usual deterministic linear amplification mentioned above [9, 10]. Experiments of amplifying coherent states noiselessly have already been demonstrated [11, 12]. This apparently imposes a significant drawback to the three-stage protocol. Nevertheless, it should be noted that the probabilistic nature of the amplification means that Eve's bit rate will further decrease. More importantly, the implementations of the amplification of coherent states operate with high fidelity only when the mean photon number after the gain is around unity [11, 12]. The amplification attack works well essentially for very weak coherent states but not for the regime of $N > 1$ that we consider in our protocol. The distortion of the quantum states at larger N introduces noise to the determination of the polarization. Further work is needed to quantify the effects of the amplification attack to the security of the protocol.

Another issue related to the amplification attack on the three-stage protocol is that, in the actual implementation, the polarization rotations U_A and U_A nevertheless have to be confined to a finite set because of the noise and stability of the

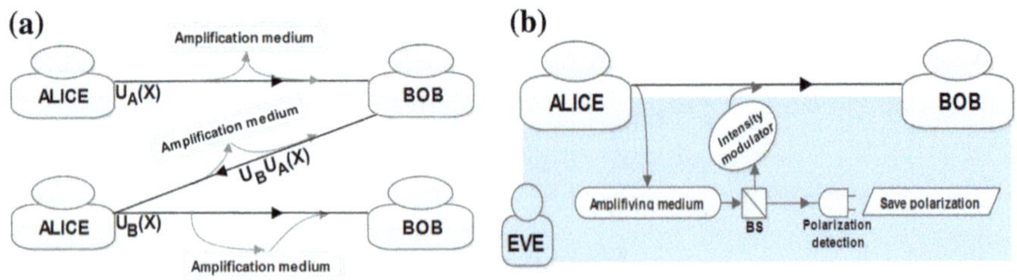

Fig. 8.5 **a** Diagram of an amplification attack on the three-stage protocol **b** diagram of Eve's amplifying medium

experimental setup. Such limitation may open up the unambiguous state discrimination (USD) attack [13, 14]. Fortunately, the polarization rotations are local information that is secret to Alice and Bob independently; they can change their sets of the rotations frequently without disclosing their actions. This results in an extremely large set of the polarization rotations and effectively mitigates the threat of the USD attack, which requires that the number of photons needed must be greater than or equal to the number of polarization states in the middle of the three-stage protocol.

8.5 Security and Key Rate Efficiency

With error correction and privacy amplification, the expression for the secret key rate extractable using one-way classical postprocessing is [15]

$$K = R[I(A:B) - \min(I_{EA}, I_{EB})], \tag{8.6}$$

where R is the raw key rate, $I(A:B)$ is the mutual information between Alice and Bob, and I_{EA} and I_{EB} are Eve's information about the raw key of Alice and Bob respectively. We consider the case when $H(A) = H(B) = 1$ and $H(A|B) = H(B|A) = h(Q)$, where $h(Q)$ is the binary entropy function and Q is quantum bit error rate (QBER). For the three-stage protocol, the raw key rate is given by the total bit that Bob measured minus the bits for authentication.

Assuming the error correction is carried out perfectly and using a very conservative estimate for the PNS/IR attack mentioned in Sect. 8.4 with mutual information $I(E:A) = 1 - h(P_e(N, tN))$, the secure key rate then becomes

$$K = R[(1-f)\, h(P_e(N, tN)) - h(Q)], \tag{8.7}$$

where f is the fraction of the MIM attacks launched by Eve, which is estimated by the ratio of the measured authentication error probability difference and the expected measured authentication error probability difference, i.e.,

$$f = \frac{P_e^{\text{Auth,measured}}(t,N) - P_e^{\text{Auth,norm}}(tN)}{P_e^{\text{Auth,MIM}}(t,N) - P_e^{\text{Auth,norm}}(tN)}. \tag{8.8}$$

The threshold for the QBER is then determined by the condition $K > 0$ for some given $f < 1$ and N.

A potentially significant drawback of the three-stage protocol compared to other QKD protocols is that it requires multiple quantum communications between Alice and Bob, effectively increases the photon loss of the channel. On the other hand, the multiple-photon resilient nature of the protocol allows a larger mean photon number to start with. As an estimate, we consider the ratio of the raw bit rates between the three stage protocol and the weak-coherent state BB'84 with mean photon number 0.5. The ratio is given by

$$E = \frac{1 - e^{-Nt(3l)}}{1 - e^{-0.5t(l)}}, \tag{8.9}$$

8.6 Summary

This chapter has given a detailed security analysis to a form of quantum cryptography protocol, the three-stage multi-photon quantum cryptography system, using coherent states to encode information [16]. It is important to note that a three-stage protocol can be used as a quantum communication protocol if authentication between Alice and Bob is established before the onset of the protocol. In case this condition is not met, the three-stage protocol may be used as a QKD protocol and operate as Sect. 8.5 of this chapter has described. In particular, this chapter has showed that the three-stage protocol is resilient to the photon number splitting attack, the intercept-resend attack, and the man-in-the-middle attack with certain error probability thresholds. In addition, this chapter has obtained the secure key rate in terms of the error probabilities under the attacks considered. Importantly, it has been found that the mean photon number of the coherent states can practically be larger than 1, in contrast to most current QKD protocols in which weak coherent pulses are considered. The multi-photon multi-stage QKD scheme presented does not require pre-sharing of a key between the legitimate users like the Y00 protocol. Hence it can be used to complement such multi-photon quantum communication protocols. This chapter has also discussed the amplification and unambiguous state discrimination attacks and argued that such attacks do not impose significant threat to the multi-stage protocol.

References

1. Barbosa, G. A., et al. (2003). Secure communication using mesoscopic coherent states. *Physical Review Letters, 90*(22), 227901.
2. Yuen, H. P. (2009). Key generation: Foundations and a new quantum approach. *IEEE Journal of Selected Topics in Quantum Electronics, 15*(6), 1630–1645.
3. Zhao, Y., Qi, B., & Lo, H.-K. (2007). Experimental quantum key distribution with active phase randomization. *Applied Physics Letters, 90*(4), 044106.
4. Bagan, E., Monras, A., & Munoz-Tapia, R. (2005). Comprehensive analysis of quantum pure-state estimation for two-level systems. *Physical Review A, 71*(6), 062318.
5. Lvovsky, A. I., Sanders, B. C., & Tittel, W. (2009). Optical quantum memory. *Nature Photonics, 3*(12), 706–714.
6. Sangouard, N., et al. (2011). Quantum repeaters based on atomic ensembles and linear optics. *Reviews of Modern Physics, 83*(1), 33.

7. Wu, H., et al. (2013). Polarization-independent slow light in annular photonic crystals. *Applied Physics Letters, 102*(14), 141112.
8. Haus, H., & Mullen, J. (1962). Quantum noise in linear amplifiers. *Physical Review, 128*(5), 2407.
9. Ralph, T., & Lund, A. (2008). *Nondeterministic noiseless linear amplification of quantum systems*. arXiv preprint arXiv:0809.0326.
10. Pandey, S., et al. (2013). Quantum limits on probabilistic amplifiers. *Physical Review A, 88*(3), 033852.
11. Xiang, G.-Y., et al. (2010). Heralded noiseless linear amplification and distillation of entanglement. *Nature Photonics, 4*(5), 316–319.
12. Zavatta, A., Fiurášek, J., & Bellini, M. (2011). A high-fidelity noiseless amplifier for quantum light states. *Nature Photonics, 5*(1), 52–60.
13. Van Enk, S. (2002). Unambiguous state discrimination of coherent states with linear optics: Application to quantum cryptography. *Physical Review A, 66*(4), 042313.
14. Becerra, F., Fan, J., & Migdall, A. (2013) Implementation of generalized quantum measurements for unambiguous discrimination of multiple non-orthogonal coherent states. *Nature Communications, 4*.
15. Scarani, V., et al. (2009). The security of practical quantum key distribution. *Reviews of Modern Physics, 81*(3), 1301.
16. Chan, K. W. C., El Rifai, M., Verma, P. K., Kak, S., Chen, Y. (2015). Multi-photon quantum key distribution based on double-lock encryption. *International Journal on Cryptography and Information Security (IJCIS), 5*(3/4).

Chapter 9
Application of the Multi-stage Protocol in IEEE 802.11i

This chapter extends the application space of the multi-stage multi-photon protocol to wireless communication. In particular, it examines the viability of using the multi-stage multi-photon protocol for secure key distribution in the IEEE 802.11i protocol. Parts of this chapter are based on the authors' work previously reported in [1].

9.1 Introduction

The dynamics of the work force environment has evolved substantively due to the introduction of wireless communication. Professionals can work without being tied to a certain location. In addition to the advantages of mobility, wireless networks gained popularity due to various other benefits such as convenience, productivity, and expandability. While wireless networks are becoming more popular day after day, related security issues are expanding in unison. Due to the nature of wireless communication, snooping and modifying a transmitted wireless signal is easier compared to its wired counterpart.

The focus of this chapter is to present a model of integration of the multi-stage quantum cryptography protocol into the IEEE 802.11 wireless communication standard. It proposes a method to integrate the three-stage quantum cryptography protocol and its variants into the key distribution scheme of the IEEE 802.11 standard. Integrating the three-stage protocol with the IEEE 802.11 standard offers several benefits compared to its single-photon counterparts. These benefits include enhanced data rates and longer distances due to its ability to operate in the multi-photon domain. Integrating the multi-stage protocol and the three-stage protocol into the IEEE 802.11 network will provide such networks with a level of security offered by the three-stage protocol in the wireless domain.

P. K. Verma et al., *Multi-photon Quantum Secure Communication*,
Signals and Communication Technology,
https://doi.org/10.1007/978-981-10-8618-2_9

9.2 IEEE 802.11i

The IEEE 802.11i standard defines a Robust Security Network Association (RSNA) based on IEEE 802.1X [2] authentication. RSNA is defined to provide better authentication and confidentiality in 802.11 networks compared to that of WEP. Three entities are involved in the authentication process: the Supplicant, the Authenticator, and the Authentication Server. In general, a supplicant and an authenticator have successful authentication after they verify each other's identity and generate a secret key that can be used in subsequent data transmissions.

The process of authentication in the IEEE 802.11i consists of handshakes between the authenticator and the authentication server, between the supplicant and the authentication server, and between the supplicant and the authenticator. The handshakes used in the process of authentication result in the generation of a common secret key called the Master Session Key (MSK). The MSK key is shared between the supplicant and the authentication server and is used by the supplicant to derive a Pairwise Master Key (PMK). The authenticator derives the same PMK using the Authentication, Authorization and Accounting (AAA) key material on the server side that is securely transferred to it. In other cases, the supplicant and the authenticator may be configured using a static Pre-Shared Key (PSK) to generate the PMK. In some other cases such as re-association, a cached PMK can be used in order to reduce the overhead employed on the authentication server when the same user undergoes repeated authentication processes.

After the establishment of a PMK, a four-way handshake protocol is executed. The four-way handshake key management protocol confirms the existence of the PMK, and the protocol generates a Pairwise Transient Key (PTK) for each subsequent session, synchronizes its installation into the MAC, and transfers the Group Transient Key (GTK) from the authenticator to the supplicants. Successful implementation of the four-way handshake means that a secure communication channel can be constructed between the authenticator and the supplicant during following data transmissions. The same PMK can be used in different four-way handshakes.

9.2.1 *The Four-Way Handshake*

Once the key between the authenticator and supplicant has been shared, the four-way handshake process will be started either by the authenticator itself or after a request from the supplicant. The message exchange is shown in Fig. 9.1.

The EAP exchange performed prior to the four-way handshake provides the shared PMK secret key. This key should be exposed as little as possible since it is intended to last for the complete session. The four-way handshake is actually used to establish another key called the PTK which is generated by concatenating the following attributes: PMK, ANonce, SNonce, AP MAC address, and STA MAC address. The product is then passed through a cryptographic hash function.

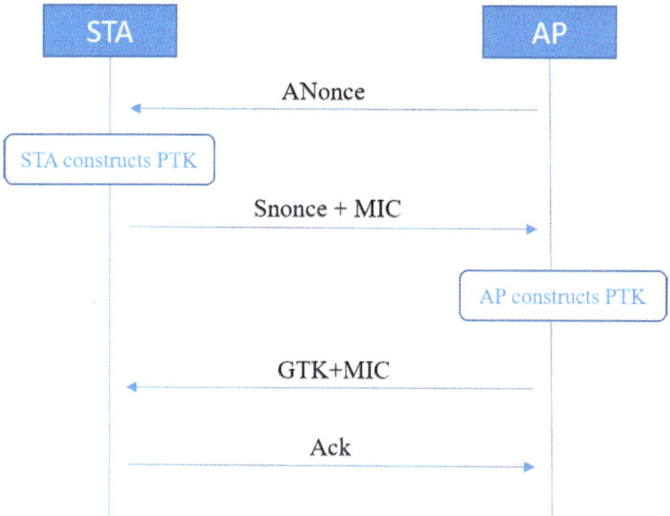

Fig. 9.1 Four-way handshake message exchange between an access point AP and a station STA

The messages exchanged during the handshake represented in Fig. 9.1 are explained below [3]:

1. The Access Point AP sends a nonce-value to the Station STA (ANonce). The STA now has all the elements needed to construct the PTK.
2. The STA sends its own nonce-value (SNonce) to the AP together with a MIC (message integrity code).
3. The AP sends the GTK and a sequence number together with another MIC to be used in the next multicast or broadcast frame. In this way, the STA can perform basic replay detection.
4. The STA sends an acknowledgement to the AP.

The PTK is then subdivided into three main parts: Key Confirmation Key (KCK), Key Encryption Key (KEK), and Temporal Key (TK). Figure 9.2 depicts the pairwise key hierarchy:

Fig. 9.2 Pairwise key hierarchy

Some of the flaws of four-way handshake are explored in [3–7]. In addition, the tool Aircrack-ng is a 802.11 WEP and WPA-PSK keys cracking program that can recover keys once enough data packets have been captured. To prevent such attacks and provide secure communication, authors of [8] have proposed the integration of the BB84 protocol into the four-way handshake. The four-way handshake with integrated QKD is called the quantum handshake. Quantum handshake is discussed in the following Section IKI.

9.3 Integration of QKD for Key Distribution in IEEE 802.11i

IEEE 802.11 wireless local area network is a suitable candidate to be used in conjunction with quantum key distribution protocols. That is due to its limited coverage area of around 200 m. Wi-Fi networks can offer a line-of-sight path, which is a major requirement for QKD. Furthermore, the sum of users of Wi-Fi is increasing significantly. Currently, all laptops are equipped with a Wi-Fi interface by default. This indicates that Wi-Fi networks must provide secured connections for the users of their services. As discussed in the previous chapters, QKD has the ability of providing unconditional security, thus it can offer Wi-Fi networks with the highest level of security that can be achieved.

The mode of integrating QKD into the key distribution process in Wi-Fi networks has been done with minimal modifications to the existing IEEE 802.11 and IEEE 802.1X protocols. The communication process implementing the four-way handshake and the initial exchange of capability parameters are the only key areas impacted when integrating QKD in the IEEE 802.11i standard.

In IEEE 802.11 standard, during the first flows of the communication, association between STA and AP is established. These initial flows involve Beacon, Probe Request and Probe Response frames. The beacon frame enables stations to establish and maintain communications in an arranged method. STA and AP use the initial flow messages to agree on the set of parameters used for following communications. When QKD is integrated in the process, it is necessary that the STA and the AP agree on all QKD related parameters at the beginning. These key parameters include: QKD protocol, reconciliation method, hash function used in privacy amplification, and quantum transmission rate. QKD parameters should be negotiated via Beacon, Probe Request, and Probe Response messages. Since the Beacon and the Probe Response frames contain Capability information fields, AP will inform prospective STAs of its capabilities of supporting QKD via the Capability information field. In addition STA, on its side, informs a prospective AP of its capabilities of supporting QKD via the Request Information and Vendor Specific fields. The information capability elements are used to transfer all the QKD specifics. It is worth noting that when any STA or AP does not support QKD, it can still proceed with the original IEEE 802.11i protocol.

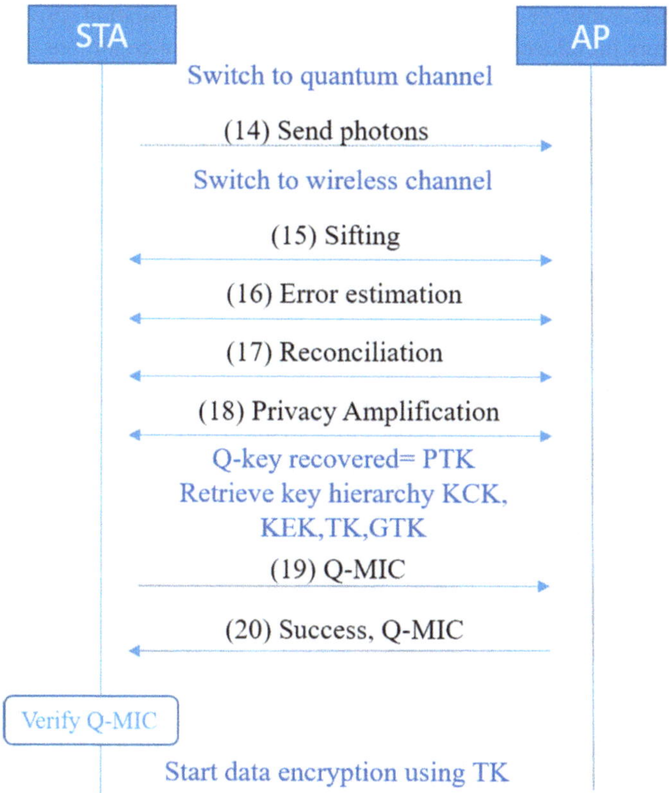

Fig. 9.3 Quantum handshake procedure

The protocol proposed to integrate the QKD process in the IEEE 802.11i networks is shown in Fig. 9.3 and described as follows [8, 9]:

The overall communication is hybrid and is done via two channels: Quantum Channel and Classical Channel. The quantum channel is used to transmit polarized photons that represent the key bits only at the beginning of the quantum handshake. After the PMK has been shared, both STA and AP switch to the quantum channel. On the other hand, the classical channel is then used later on to retrieve the final secret key by implementing the proposed quantum handshake protocol.

After EAP Key message, both STA and AP start the QKD protocol by switching to the quantum channel. In flow (14) of Fig. 9.3 the STA sends a sequence of polarized photons representing the random key that STA intends to share with the AP. Flow (14) corresponds to the first three steps of the BB84 protocol. The number of photons that an STA sends depends on the length of the PTK, the QKD protocol used, and the privacy amplification algorithm. Sufficient number of photons should be sent, since during the communication over the classical channel, some of these bits conveyed over the quantum channel get discarded. Bits are eliminated due to:

sifting, dark counts of apparatus, error estimation, reconciliation, privacy amplification, errors introduced due to noise, and environmental conditions.

When the AP receives the photons it effects step 4 of the BB84 protocol. Then AP and STA switch back to the classical channel. From this point forwards, the key recovery process based on the BB84 protocol takes place. Using EAPOL frames for communication, AP and STA remove the bits that are in error and end up with identical and secured key. During flow (15) of Fig. 9.3 they effect the sifting process of the BB84 protocol. Using flow (16) of Fig. 9.3 STA and AP effect the step of error estimation of the BB84 protocol.

Using flow (17) of Fig. 9.3, STA and AP effect the most crucial phase of the quantum handshake corresponding to the reconciliation process. This is step in the BB84 protocol is done to correct the errors in their keys and recover identical keys. The error correction algorithm that the STA and AP will follow at this stage has already been decided by the Beacon, Probe Request and Probe Response messages at the beginning of the communication. The STA and AP implement the reconciliation protocol agreed on via EAPOL message communications. At the end of the reconciliation phase both STA and AP have identical copies of an error free key. The next step is the privacy amplification implemented using flow (18) of Fig. 9.3.

The final key recovery process of the quantum handshake involves removing some bits from their position in the raw key. In this case, the quantum transmission must ensure sending sufficient number of photons in order to recover final key that contains a number of bits at least equal or greater than that of PMK. Any extra bits of the final key will be removed so that it will have same length as PTK.

Using flow (19) of Fig. 9.3, STA sends Q-MIC to AP. AP calculates its own version of Q-MIC and upon receiving Q-MIC from STA, it compares it with the Q-MIC version it has calculated. If they match, the STA is authenticated and AP will send a success message using flow 20. Since the PTK is shared by performing QKD, PTK is unconditionally secure. Thus the rest of the key structure of 802.11 comprising KCK, KEK and TK inherit the same level of security as PTK.

9.3.1 Disadvantages of the Approach Described to Integrate QKD into IEEE 802.11i

Integrating the BB84 into the four-way handshake of the IEEE 802.11i protocol has several limitations:

(1) The AP and STA might not be able to have a direct line of sight even in an environment where an AP and STA are only distant by few meters. In such cases the solution is to use the original IEEE 802.11i protocol.
(2) Single photon generator and single photon detectors are required on both AP and STA sides. Due to the Heisenberg uncertainty principle, we cannot have any device that can reliably generate a single photon per time slot. Current QKD systems rely on attenuated laser pulse that may generate less than one photon per

time slot, thus all but guaranteeing that most time slots will be either empty or carry no more than a single photon per pulse. Therefore, an STA will need much more time to send the required number of photons, and in addition photon pulses containing more than one photon per pulse will make the system vulnerable to the photon number splitting attacks. On the other hand, avalanche photodetectors are used at the AP side. APDs require a cooling temperature of −50 °C to reduce the dark counts. Also in a wireless environment, vulnerability of a signal containing a single photon to environmental noise will be higher than its wired counterpart. This is why it is important to look into solutions where different QKD multi photon tolerant protocols can be used.

(3) The requirement of a single photon per pulse for BB84 limits the distance and the data rate at which the PTK can be shared. Thus solutions provided by using multi-photon protocol is proposed in the next chapter of this report.

(4) Furthermore, the BB84 has several steps after the quantum transmission process. These steps include: sifting, error estimation, error correction, and privacy amplification. Compared to the cases using a multi-photon tolerant protocol, these steps are seen as an overhead associated with the BB84 protocol execution in addition to the associated cost of implementing them, where a shorter key will be derived. Furthermore, multiple agents should be present to implement these steps. A multi-agent based approach with a lesser number of needed agents will be discussed later in this chapter.

The first proposed multi-photon tolerant protocol is the three-stage protocol. It was first proposed by Subhash Kak in 2006, this protocol can obviate some of the known limitations of the BB84 protocol such as the distance, data rate, and single photon requirement. One can think of the three-stage protocol as a candidate to replace the BB84 in the quantum handshake discussed in the precious chapter. The three-stage protocol described in Chap. 5 has two other variants: the three-stage protocol using an initialization vector and implementing a chaining mode between the keys agreed on (three-stage protocol using four variables) discussed in Chap. 5, and the single-stage protocol where an initialization vector is used as the set of initial transformations to be applied on the message transmitted. In the case of the single-stage protocol one transmission only is needed to convey a message. The next part of this chapter will present a discussion of the integration the three-stage protocol and its variants into the IEEE 802.11 quantum handshake.

9.4 Hybrid Three-Stage Protocol

The three-stage protocol is composed of three consecutive messages that require a line-of-sight between the STA and the AP. One can think of implementing a hybrid approach were a reduced number of transmission requiring a line of sight can be used.

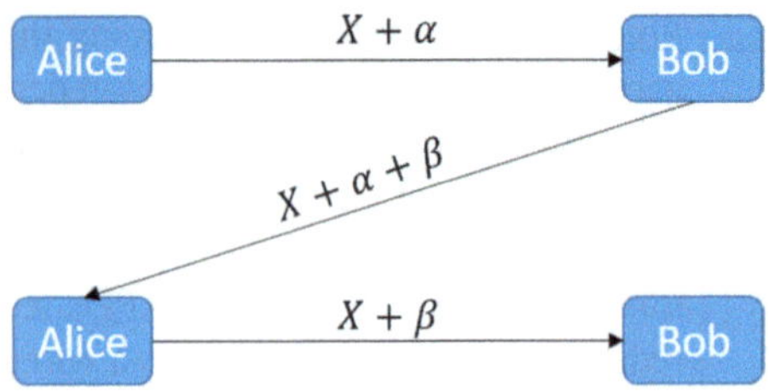

Fig. 9.4 The three-stage protocol

The three-stage protocol is shown in Fig. 9.4. The three-stage protocol proceeds as follows:

Step 1: Alice sends the polarization value of the first transmission using the classical channel. The value of the polarization is equal to $X + \alpha$, where X is the value of the corresponding bit of the key (90° for bit value 1 and for bit value 0°) and α is the rotation applied by Alice.

Step 2: Bob receives the value of the polarization, generates a number of photon having the corresponding polarization and applies his rotation transformation and send using the quantum channel the set of photons having a polarization of $X + \alpha + \beta$ back to Alice, where β is the polarization rotation applied by Bob.

Step 3: Alice removes her transformation and sends using the quantum channel the set of photons of polarization equal to $X + \beta$.

Step 4: Bob will receive the set of photons from Alice removes his transformation and measure the initial value of X.

9.4.1 Quantum Handshake Using the Three-Stage Protocol

If we replace the BB84 protocol in the IEEE 802.11i protocol (described in the previous section) by the three-stage protocol the only part affected will be the quantum handshake. The quantum handshake using the three-stage protocol is as shown in Fig. 9.5.

As shown in Fig. 9.5 the quantum handshake using the three-stage protocol requires less message flows between the AP and the STA.

9.4.2 Quantum Handshake Using the Four-Variable Three-Stage Protocol

The three-stage protocol using four-variable is described in [10]. In this protocol an initialization vector *IV* is used to enhance the security of the three-stage protocol. Then a chaining mode is implemented in order to update the value of the

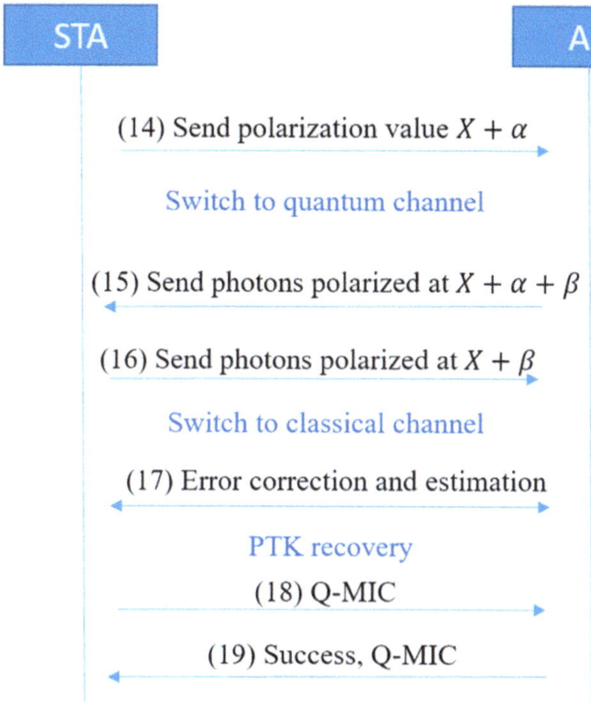

Fig. 9.5 Quantum handshake using the three-stage protocol

polarization vector. In this case the PMK can be used as an initialization vector and for later transmission the PTK can be used to update the value of the initialization vector. The three-stage protocol using four variables was described in details in Chap. 6. The integration of this variation of the three-stage protocol into the quantum handshake is shown in Fig. 9.6.

9.4.3 Quantum Handshake Using the Single-Stage Protocol

In the single-stage protocol, it is assumed that Alice and Bob use transformation known to both Alice and Bob (in our case STA and AP). Thus, rather than using three transmissions to convey the key STA and AP can share the key using one transmission. Prior to the communication Alice and Bob should have a set of transformations that they will be applying to the messages exchanged among them.

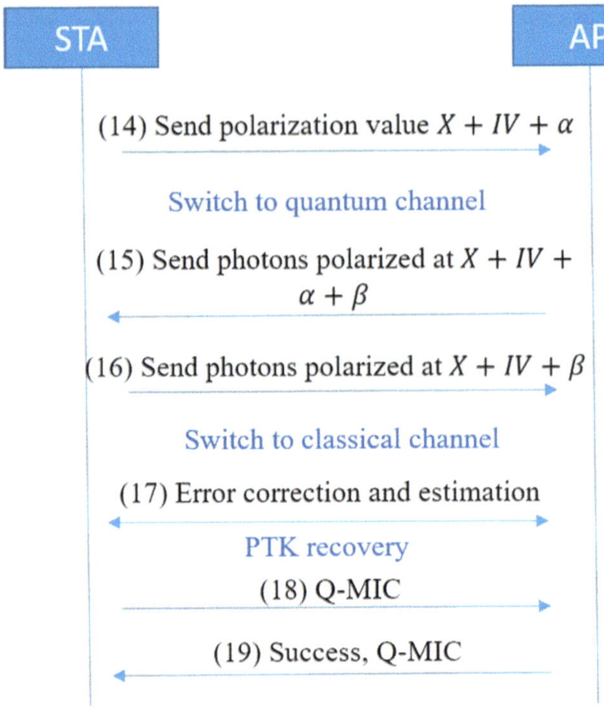

Fig. 9.6 Quantum handshake using the four variable three-stage protocol

This set of transformation in this case can be derived from the PMK and updated according to the value of the shared PTK.

The steps of the single-stage protocol are as follows [11]:

Step 1: Alice apply her transformation on the message and sends $U_A(X)$ to Bob.
Step 2: Bob removes the transformation applied by Alice, by applying its transpose complex conjugate since he has a prior knowledge of this transformation.

Figure 9.7 shows the quantum handshake of the IEEE 802.11i protocol using the one-stage protocol.

9.4.4 *Hardware Implementation*

Till date there are no commercially available Wi-Fi specific wireless devices that support quantum transmission. Thus the quantum transmission in [8] has been established as a separate project aligned to an existing QKD research [12–14] where the quantum transmissions have been practically implemented over free space. The setup between AP and STA is shown in Fig. 9.8.

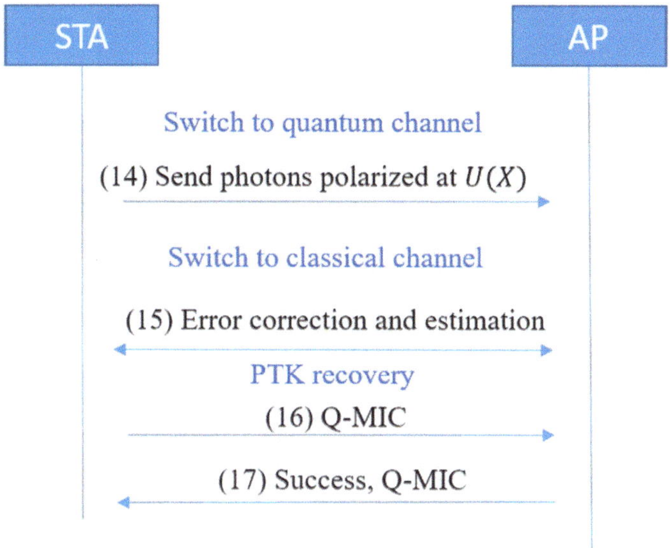

Fig. 9.7 The quantum handshake of the IEEE 802.11i using the single-stage protocol

9.5 Software Implementation

The implementation was done using multi agent system. An agent is a sophisticated computer program capable of acting autonomously to accomplish tasks on behalf of its users, across open and distributed environments [15, 16]. It has the following characteristics: autonomy, mobility, rationality, reactivity, inferential capability, pro-activeness, and social ability, etc.

Together multiple agents form a Multi-agent System (MAS) offer several advantages over a centralized approach. They can distribute computational resources and capabilities across a network. Multiple agents enhance overall system performance, efficiency, reliability, extensibility, robustness, maintainability, responsiveness, flexibility, and reuse.

9.5.1 Multi-agent Approach in BB84

The QKD based IEEE 802.11i using BB84 can be represented using the agents shown in Fig. 9.9.

The functionalities of each agent can be described as follows [8]:

802.11 Agent: This agent performs the 802.11 Association and Authentication.

802.1X Agent: This agent carries out the 802.1X authentication. It is also capable of making decisions on suspicious messages from adversaries similar to what 802.11 Agent does.

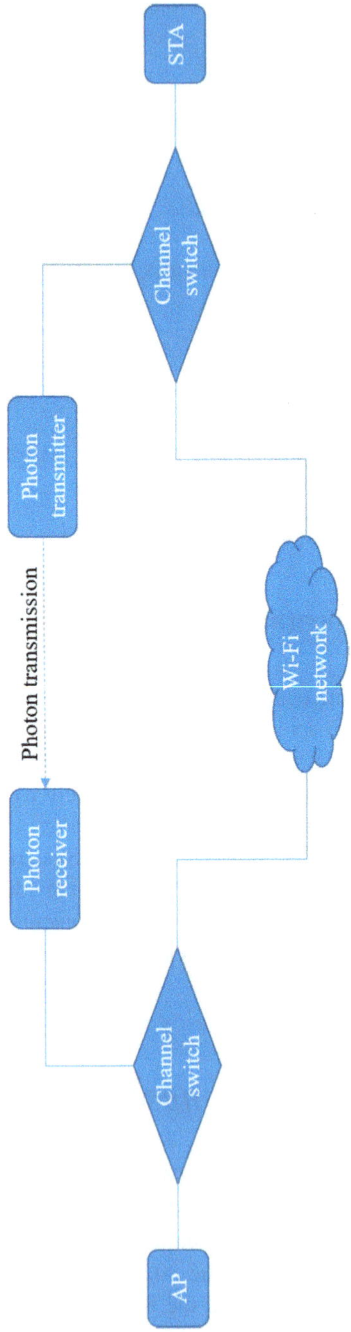

Fig. 9.8 Implementation setup of the IEEE 802.11i integrated with QKD

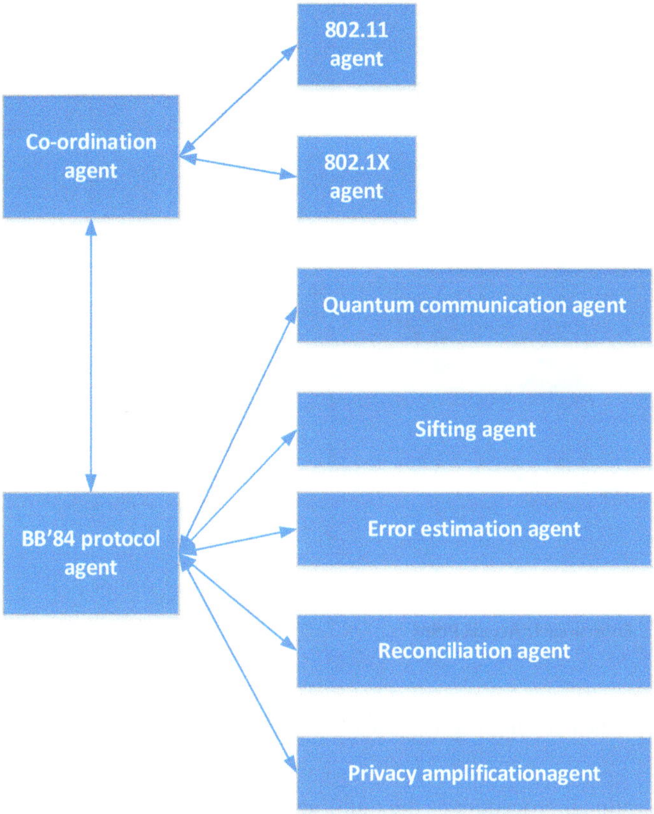

Fig. 9.9 Multi-agent approach to BB84 in IEEE 802.11i. *Source* [8]

BB84 Agent: This agent is to act as the coordinating agent to execute the BB84 QKD protocol. It communicates with 4 other agents to execute the BB84 protocol.

Sifting Agent: This agent effectuates the sifting stage of the BB84 protocol.

Error Estimation Agent: This agent verifies if the error level of the quantum transmission is acceptable or not.

Reconciliation Agent: This agent executes the reconciliation process of the BB84 protocol to remove incorrect bits and obtain an error free key at either end.

Privacy Amplification Agent: This agent effectuates the Privacy amplification of the BB84 protocol. The key obtained by this agent is the "unconditionally secured" PTK key used to derive the rest of the key hierarchy.

Coordination Agent: The coordination agent communicates with all other agents and assures that monitoring efforts and management of internal requests with other agents are handled consistently within a specific transmission. In other words, the coordination agent offers an outline of all communications performed between agents.

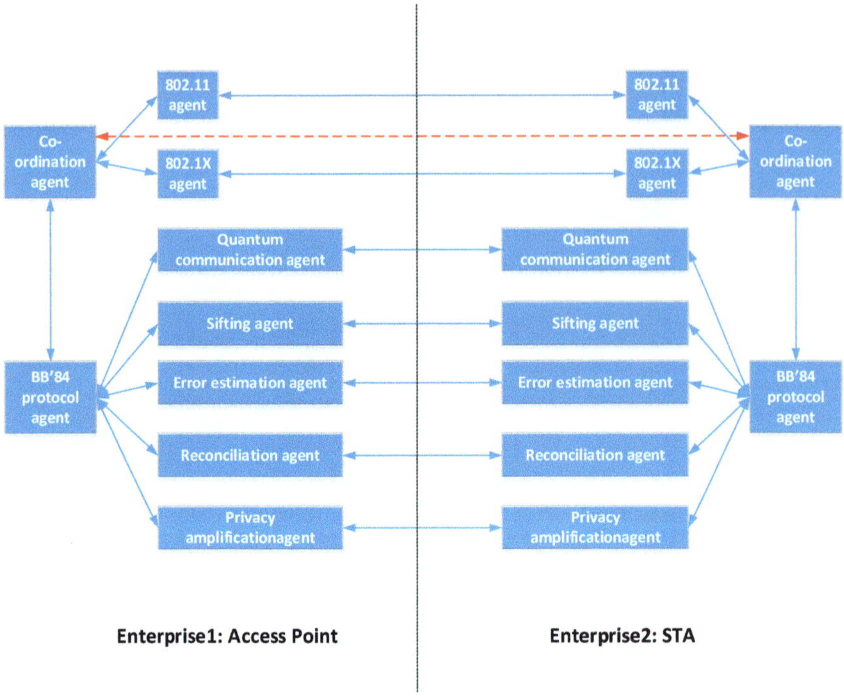

Fig. 9.10 Operation of a multi-agent approach

The operational procedure of the multi-agent approach is shown in the Fig. 9.10. Each time a new STA enters into the network, AP creates an enterprise corresponding to the new STA in order to facilitate the communication. On the other hand the STA also creates an instance of this enterprise. Since the STA communicates with only one AP at a time, it will only have one enterprise, however, the AP's side will contain many enterprises according to how many STA it is dealing with. The two enterprises that make and the communication procedure are shown in Fig. 9.9.

9.5.2 Multi-agent Approach in Multi-photon Tolerant Protocols

One can use a multi-agent approach to implement the three-stage protocol and its variant in an IEEE 802.11i network. Compared to its BB84 counterpart, such implementation will need less agents to perform the actions required in order to share a PTK between AP and STA. The agents are shown in Fig. 9.11.

As we can see the three-stage protocol and its variants doesn't need the presence of: sifting agent, reconciliation agent, and privacy amplification agent. On the other

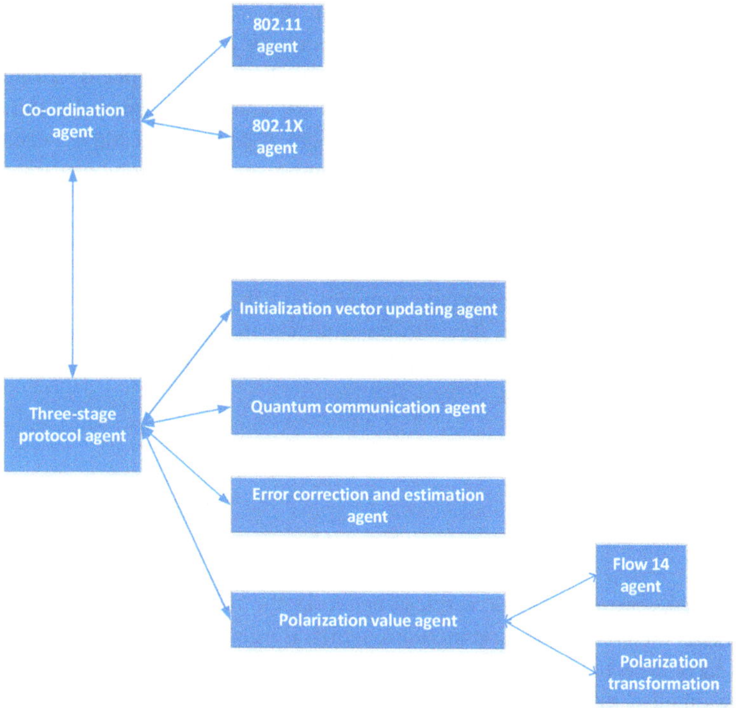

Fig. 9.11 Agents used for the three-stage (and its variants)

hand, it is useful for the software implementation of the three-stage protocol and its variants to have an initialization vector updating agent that will compute the value of the new initialization vector to be used at the next session. In addition to a polarization value agent, this agent have two wrapper agents the first is for sending the value of the polarization of the flow 14 of Figs. 9.5, 9.6 and 9.7, the second is for applying the polarization transformation on each side of the communication. It is worth noting that the polarization value agent of the AP and STA need not to communicate at any point in the session.

9.5.3 Analysis of the Quantum Handshake Using Three-Stage Protocol and Its Variants

Compared to the quantum handshake using BB84, quantum handshake using three-stage and its variants offers the following advantages:

(1) On the positive side, the three-stage protocol and its variants are multi-photon tolerant protocols. Thus, AP and STA do not need to be equipped with single

photon generators and detectors. However, the proposed approach requires the presence of a high speed polarization modulator at both AP and STA sides.

(2) On the positive side, the three-stage protocol offers a higher data rate and a longer communication distance between AP and STA. Since the three-stage protocol is a multi-photon tolerant protocol it can offer a longer distance of communication since a light pulse containing several photons can travel for a longer time. Furthermore, the three-stage protocol doesn't require sifting thus the data rate will be higher compared to that of the BB84 and one can have a better assumption of how many bits should be sent to retrieve the final key.

(3) Compared to BB84, three-stage protocol and three-stage protocol using four variable will require two flows to be sent over the quantum channel (flows 15 and 16 of Figs. 9.5, 9.6 and 9.7). However, the single-stage protocol does not impose this requirement.

9.6 Summary

Security risks are inherent in wireless network; the major source of risks in such networks is the communication medium, which is open to intruders. Till date, many efforts have been put into address security issues in wireless networks. This chapter has presented a method to provide wireless networks with a security level comparable to that of QKD. It has discussed the integration of the BB84 protocol into the four-way handshake of the IEEE 802.11i protocol along with its shortcomings. On the positive side, quantum handshake provides the Wi-Fi networks with unconditional security. On the other hand, quantum handshake using the BB84 imposes several limitations such as short distances, low data rates etc. This chapter has proposed the integration of multi-stage protocols into the four-way handshake of IEEE 802.11i [17]. Multi-stage protocols offer several advantages compared to its BB84 counterpart. The mode of operation of the proposed quantum handshake (using multi-stage protocol) has been discussed along with its multi-agent implementation approach. Providing the four-way handshake with the capability of implementing the multi-stage protocol offers a quantum level of security while achieving higher data rates and longer distances. The only disadvantage of such an approach is the requirement of at least a single line-of-sight between AP and STA. Furthermore, one can think of implementing QKD in the key distribution of Wi-Max networks [1], since such networks use key hierarchy similar to that in IEEE 802.11i.

References

1. Nomula, R., Rifai, M. E., & Verma, P. (n.d.) Multi-photon tolerant protocols for quantum secure communication in wireless standards. *International Journal of Security and Networks, 11* (in press).
2. Society, I. C. 802.1X. In *IEEE Standard for Local and metropolitan area networks Port-Based Network Access Control*. NY, USA: IEEE.
3. He, C., & Mitchell, J. C. (2004). Analysis of the 802.11i 4-way handshake. In *Proceedings of the 3rd ACM Workshop on Wireless security*, ACM.
4. He, C., & Mitchell, J. C. (2004). Message attack on the 4-Way handshake. In *Submissions to IEEE.*
5. De Rango, F., Lentini, D. C., & Marano, S. (2006). Static and dynamic 4-way handshake solutions to avoid denial of service attack in Wi-Fi protected access and IEEE 802.11i. *EURASIP Journal on Wireless Communications and Networking, 2006*(2), 73.
6. Mitchell, C. H. J. C. (2005) Security analysis and improvements for IEEE 802.11i. In *The 12th Annual Network and Distributed System Security Symposium (NDSS'05)*, Citeseer.
7. Bai, Z., & Bai, Y. (2009) 4-way handshake solutions to avoid denial of service attack in ultra wideband networks. In *Intelligent Information Technology Application, 2009. IITA 2009. Third International Symposium on*, IEEE.
8. Wijesekera, S. (2011) *Quantum cryptography for secure communication in IEEE 802.11 wireless networks.* Canberra: University of Canberra.
9. Nguyen, T. M. T., Sfaxi, M. A. & Ghernaouti-Hélie, S. (2006). Integration of quantum cryptography in 802.11 networks. In *Availability, Reliability and Security, 2006. ARES 2006. The First International Conference on*, IEEE.
10. El Rifai, M., & Verma, P. K. (2013). *An algorithmic approach to securing the three-stage quantum cryptography protocol.* In *Trust, Security and Privacy in Computing and Communications (TrustCom), 2013 12th IEEE International Conference on*, IEEE.
11. Thomas, J. H. (2007) *Variations on Kak's three stage quantum cryptography protocol.* arXiv preprint arXiv:0706.2888.
12. Ganeshkumar, G., et al. (1999) *The university of Canberra quantum key distribution testbed.*
13. Edwards, P., & Lynam, P. (2002) *The University of Canberra–Telstra tower free-space quantum key distribution testbed.* ITEE Society Monitor (March 2002).
14. Edwards, P. J. *The University of Canberra–Telstra tower quantum crypto-key telecommunications link, advanced telecommunications and electronics research centre.*
15. Wijesekera, S., Huang, X., & Sharma, D. (2009) Multi-agent based approach for quantum key distribution in WiFi networks. In *Agent and Multi-Agent Systems: Technologies and Applications* (pp. 293–303). Berlin: Springer.
16. Weiss, G. (1999). *Multiagent systems: A modern approach to distributed artificial intelligence*. Cambridge: MIT press.
17. El Rifai, M., & Verma, P. K. (2015) Quantum secure communication using a multi-photon tolerant protocol. In *Proc. SPIE 9377, Advances in Photonics of Quantum Computing, Memory, and Communication VIII, 937713* (March 4, 2015), https://doi.org/10.1117/12.2077229.

Chapter 10
Intrusion Detection on Optical Fibers

This chapter discusses an application of the polarization property of light in detecting intrusion on an optical fiber with the objective of stealing information flowing through it. Detection of intrusion, if timely accomplished, will offer an effective means to prevent information from being captured by a malicious agent. The material presented in this chapter is based on the authors' work previously published in [1].

10.1 Intrusion Detection and Encryption

The previous chapters of the book have addressed means of encrypting information using the polarization property of light. Thus, an unauthorized party can possess encrypted data but will not be able to make any sense out of it. This chapter uses the polarization property of light to prevent leakage of information in the first place. In a sense, it's a more powerful technique that could obviate the need for encryption. However, it's also limited in a variety of ways as discussed below.

First, detection of any malicious pilferage of information must happen instantly to be effective. Second, in the technique we discuss, the detection is limited to information *flowing over an optical fiber*. It does not detect pilferage of static information such as information stored in a storage medium.

Detecting pilferage of information, whether it's in a static or transit mode is, fundamentally, a difficult proposition. Figure 10.1a shows an easy detection of material theft by discovering its shortage or absence in its entirety. Information theft, shown in Fig. 10.1b, is undetectable because it can be copied indefinitely with little cost of storage or transfer.

In many ways, detection of intrusion and encryption are complementary means of protecting the integrity of information. Even if we had a foolproof way of protecting access to information by an unauthorized party, encryption technology will still have an important role in managing the legitimate use of information. As a

P. K. Verma et al., *Multi-photon Quantum Secure Communication*,
Signals and Communication Technology,
https://doi.org/10.1007/978-981-10-8618-2_10

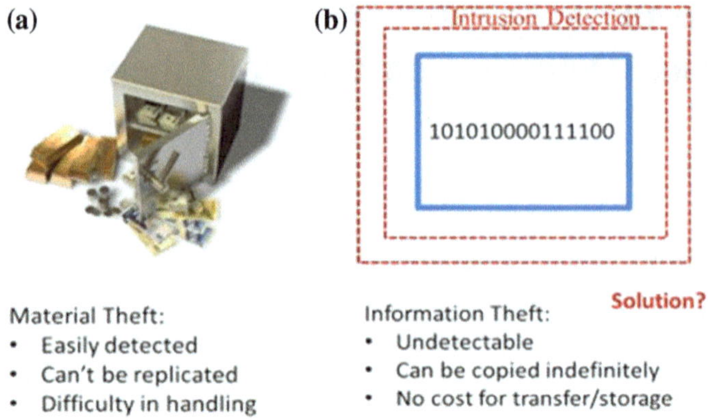

Fig. 10.1 **a** Material theft, **b** Information theft

relatively straightforward example, encryption of information by the private key of a party is an effective means of establishing the authenticity of information as having originated from the named party.

We have noted in Chap. 3 that the BB84 is, fundamentally, an intrusion detection system. The intrusion detection allows the communicating parties to abort a communication if an intrusion is suspected. This allows the parties to communicate in an intrusion-free environment and take care of the transmission errors by suitable forward error correction techniques.

This chapter explores the possibility of creating an intrusion detection system in fiber-optic cables using the polarization channel of the optical fiber. The proposed instrumentation is simple and the detection can be carried out without the burden of generating or detecting single photons.

10.2 Tapping of Optical Fibers

The physical construction of an optical fiber is shown in Fig. 10.2. As shown, it consists of a fiber core surrounded by a cladding with a lower index of refraction. An outer jacket protects the cladding and the fiber from environmental damage.

There are several ways in which information can be tapped out of an optical fiber compromising the security of information. For example, when the fiber is bent, and its radius reaches a critical angle, light leaks out of the core and it can be detected by an optical detector. The detected light can be processed to extract information being transferred over one or more channels of a wavelength division multiplexed system. As another example, the original fiber can be cut and a splitter or coupler inserted in the light path.

The countermeasures that can be deployed to protect the physical integrity of an optical fiber, and thus prevent intrusion into it, are far from being practical. For

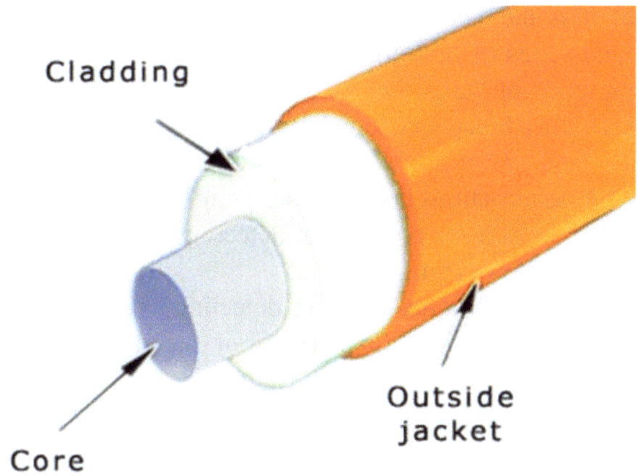

Fig. 10.2 Cross-section of an optical fiber

example, the optical fiber can be encased in concrete preventing anyone from accessing it. Alternatively, the fiber can be installed in a pressurized conduit and continuous surveillance could be maintained to ensure its integrity from any physical encroachment by measuring the pressure differential. Neither of these techniques are practical in a real-life environment.

More recently, several authors [1–3] have proposed ways to use the polarization properties of light to detect intrusion. This chapter summarizes their investigations and presents the actual instrumentation they have used to verify their approach.

10.3 Polarization Properties of Light [1]

MacDonald and Sluss [2] have investigated the properties of light as a means of insuring the integrity and security of the physical layer of a fiber optic communication link. Specifically, the authors have focused on the behavior of polarization in a single mode fiber, as it is shown to be especially sensitive to fiber geometry and to changing environmental conditions. Accordingly, we posit that the state of polarization (as represented in Stokes space) is more sensitive to environmental conditions than either the degree of polarization, or to the received power (S_0).

Recall from Chap. 2 that the degree of polarization is represented as $DOP = \sqrt{S_1^2 + S_2^2 + S_3^2}/S_0$. Given negligible changes in DOP and S_0, a change in one Stokes parameter necessitates a change in at least one other Stokes parameter. From a practical standpoint, this last case is observed only under controlled conditions. What is usually observed are changes in S_1, S_2, and S_3. *Furthermore, "small" changes over time are usually due to changing environmental conditions, while "large" changes are due to fiber perturbations.*

We regard polarization measurements as sampled points from a continuous function. In time series analysis, a common preprocessing step is to remove any trend in the data by taking the first-differences of the sampled points. The resulting time series of first-differences can often be regarded as a stationary process. The presence of "large" first-differences in the time series of first-differences is interpreted as being generated by a mechanism that is different from the generating mechanism behind "small" first-differences. We interpret the former as arising from fiber perturbations and the latter as arising from changing environmental conditions. We approach the problem of detecting fiber perturbations as a time series anomaly detection problem.

Many techniques exist for detecting anomalous events. A few include: (1) from ordinary statistics; the kurtosis of the distribution of first-differences, (2) from spatial statistics; the Hopkins metric for identifying cluster formation [4], and (3) from machine learning; one-class support vector machines (OCSVM). In this paper we employ the use of Extreme Value Theory (EVT) for detecting anomalous events.

EVT has the advantage of being able to adapt to changing environmental conditions, and encapsulating varying notions of "unusualness" (degrees of anomalous behavior). Essentially, we use EVT to characterize the first-difference behavior of unperturbed fiber so that first-differences produced by perturbations can be identified probabilistically.

Extreme value theory employs two main approaches [5]: Peak over threshold model and block maxima model. Block maxima, used in this work, uses the maximum occurring data value from each block (a contiguous set of first-differences) without regard to preset thresholds. EVT essentially addresses the following question: "What is the probability of observing some quantity that is more extreme than any quantity so far observed?" We do not explain EVT here but later make references to "sensitivity threshold." This phrase refers to the magnitude of first-differences that are likely to be observed only under rare circumstances (fiber perturbations) as determined by the EVT block maximum method. (EVT suggested threshold in terms of normalized Stokes parameters was 0.05.)

10.4 Experimental Setup

Section 3 has presented the fundamental components of the intrusion detection system. This section describes the implementation of the system. The system was tested in the Quantum Optics Lab at OU-Tulsa, where the surrounding temperature was around 65 °F. In the conducted experiment, depicted in Fig. 10.3, measurement of the state of polarization was made by Polarization analyzer: Agilent 8509C lightwave polarization analyzer, which offers high-speed, calibrated polarization measurements of optical signals and components. It measures the state of polarization, the degree of polarization and the power. Additionally, LabVIEW 10.0.1 controls the experiment and collects the measured data. The single mode optical fiber has FC connectors at both ends.

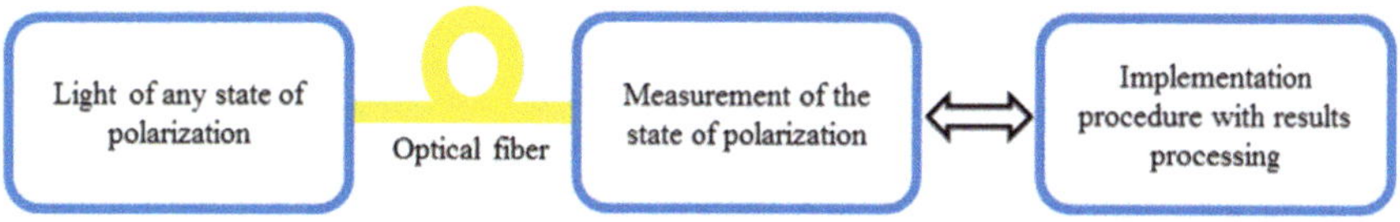

Fig. 10.3 Schematic diagram of polarization-based intrusion detection system consists of: the measured data, and optical fiber with FC connectors on both ends

The system was also tested with multimode optical fiber with FC connectors. The operating wavelength was 1550 nm. The LabVIEW communicated with the Agilent 8509C lightwave polarization analyzer. It also saved the collected measurements in a text file for further processing. The degree of polarization, power, and state of polarization of the received light were measured. The LabVIEW controller and driver generated a text file that included the following (Fig. 10.4): SeqNumber, Timestamp, LinkStatus, Power, S_1, S_2, S_3, Degree of Polarization, Heartbeat Interval, FutureUse1, FutureUse2.

- Sequence number: helps identify missing measurements.
- Timestamp: the time and date when the measurement is made. The time is in the 24-h format.
- Link status: OK, Intrusion, and Suspicious: link status depends on the first-differences of the Stokes parameters. If the first-differences of S_1, S_2, and S_3 are below the sensitivity threshold, link status is OK. If they are all above the threshold, it is intrusion. In the other cases, the link status is suspicious.
- Measured raw data: power of the received signal, the state of polarization in terms of the Stokes parameters (S_1, S_2, and S_3), and the degree of polarization.

```
20,10:00:47 9/15/2014,OK,4.312181E-05mW,-.9642598,.1602537,-.1817518,.9942394,5s,futureuse1,futureuse2
21,10:00:49 9/15/2014,OK,4.312107E-05mW,-.9640558,.1594267,-.1821155,.9939751,5s,futureuse1,futureuse2
22,10:00:52 9/15/2014,OK,4.310184E-05mW,-.9638325,.1606475,-.181856,.9939076,5s,futureuse1,futureuse2
23,10:00:54 9/15/2014,OK,4.309892E-05mW,-.9639255,.1611923,-.1805002,.9938387,5s,futureuse1,futureuse2
24,10:00:56 9/15/2014,OK,4.312029E-05mW,-.9640692,.1605127,-.1813008,.994014,5s,futureuse1,futureuse2
25,10:00:58 9/15/2014,OK,4.313569E-05mW,-.9647786,.1584617,-.1806245,.99425,5s,futureuse1,futureuse2
26,10:01:01 9/15/2014,OK,4.310966E-05mW,-.9643043,.1600006,-.1807907,.9940665,5s,futureuse1,futureuse2
27,10:01:03 9/15/2014,OK,4.312015E-05mW,-.9645903,.1601306,-.1810278,.9944081,5s,futureuse1,futureuse2
28,10:01:05 9/15/2014,OK,4.311911E-05mW,-.9635097,.1608307,-.1811701,.9934989,5s,futureuse1,futureuse2
29,10:01:08 9/15/2014,OK,4.312312E-05mW,-.9643953,.1608168,-.1803,.9941974,5s,futureuse1,futureuse2
30,10:01:10 9/15/2014,OK,4.311878E-05mW,-.9637945,.1611042,-.1802603,.993654,5s,futureuse1,futureuse2
31,10:01:12 9/15/2014,OK,4.3143E-05mW,-.9640971,.1594487,-.181045,.9938232,5s,futureuse1,futureuse2
32,10:01:15 9/15/2014,OK,4.31442E-05mW,-.9645333,.1588815,-.1819154,.9943144,5s,futureuse1,futureuse2
33,10:01:17 9/15/2014,OK,4.313379E-05mW,-.963989,.1606798,-.1809573,.9939006,5s,futureuse1,futureuse2
34,10:01:19 9/15/2014,OK,4.314482E-05mW,-.9642271,.1593157,-.1807891,.9938813,5s,futureuse1,futureuse2
35,10:01:22 9/15/2014,OK,4.313351E-05mW,-.9639977,.1599554,-.1809125,.993784,5s,futureuse1,futureuse2
36,10:01:24 9/15/2014,OK,4.314826E-05mW,-.964191,.1592727,-.1808593,.9938522,5s,futureuse1,futureuse2
37,10:01:26 9/15/2014,OK,4.312597E-05mW,-.9645693,.1600896,-.1808089,.9943413,5s,futureuse1,futureuse2
38,10:01:28 9/15/2014,OK,4.312447E-05mW,-.9642321,.1597496,-.1805179,.9939066,5s,futureuse1,futureuse2
39,10:01:31 9/15/2014,OK,4.314412E-05mW,-.9638712,.1599187,-.1802871,.9935417,5s,futureuse1,futureuse2
40,10:01:33 9/15/2014,OK,4.312131E-05mW,-.9652541,.158798,-.179218,.9945105,5s,futureuse1,futureuse2
41,10:01:35 9/15/2014,OK,4.314331E-05mW,-.9640119,.1600732,-.1792498,.9935154,5s,futureuse1,futureuse2
42,10:01:38 9/15/2014,OK,4.315595E-05mW,-.9641436,.1595142,-.1800317,.9936947,5s,futureuse1,futureuse2
43,10:01:40 9/15/2014,OK,4.311647E-05mW,-.9645304,.1588167,-.1794175,.9938473,5s,futureuse1,futureuse2
44,10:01:42 9/15/2014,OK,4.311786E-05mW,-.9642872,.1599119,-.1796924,.9938364,5s,futureuse1,futureuse2
45,10:01:45 9/15/2014,OK,4.312738E-05mW,-.964204,.1589892,-.1815305,.9939418,5s,futureuse1,futureuse2
```

Fig. 10.4 Sample text file of collected measurements

10.5 Experimental Results

The intrusion detection system was tested under a variety of scenarios. The results of those tests are presented in this section. The system was tested with a 10-km unperturbed single-mode fiber. For the unperturbed fiber, Stokes parameters, degree of polarization, and power were relatively stable. The first-differences of the Stokes parameters were relatively stable as well, and did not exceed the EVT suggested sensitivity threshold. For this same unperturbed fiber, the input polarization state was modified using the Agilent 8169A polarization controller. As expected, the change in the input polarization state resulted in a significant change in the received state of polarization but with negligible differences in the degree of polarization and power values. Furthermore, the time-evolving Stokes parameters and their first-differences were relatively stable and did not exceed the sensitivity threshold. Simulation of changing environmental conditions (specifically, minor vibrational fluctuations and temperature variations) led to gradual changes in the values of the Stokes parameters. The resulting trajectory on the Poincare sphere traced out by the changing Stokes parameters tended to be more organized. However, first-differences of the Stokes parameters were relatively stable and did not exceed the sensitivity threshold. This suggests that the intrusion system is robust against changing environmental conditions, namely occasional vibration and temperature. (Vibrational stimuli were applied by affixing the fiber cable to a cell phone set on vibrate mode.) The vibration stimuli had a frequency of 180 Hz. Temperature fluctuations were induced using a heating element in close proximity to the fiber cable. The applied external heat was in the range of 1500–1800 W, which resulted in a temperature range of 70–130 °F. The system was tested with a single-mode fiber for 24 h with occasional variations in temperature, vibration, and perturbation. The resulting measurements were plotted and are shown in Fig. 10.5. The system was tested with the following alterations:

- At interval 3141–3378, external heat was applied to the fiber. This resulted in a change in the values of S_1, S_2, and S_3. However, the first-differences were stable below the sensitivity threshold.
- At interval 6235–6490, the fiber was subjected to external vibration. There was a smooth change in the values of the Stokes parameters and only the first-difference of S_3 exceeded the threshold. In both intervals, the power and the degree of polarization were stable and did not reflect the external alterations.
- At interval 10178–10203, the fiber was bent. This event can be clearly seen in the values of S_1, S_2, and S_3 that showed an abrupt change. Also, the magnitude of the first-differences for all Stokes parameters far exceeded the sensitivity threshold, suggesting a different generating process. Only negligible changes in the degree of polarization and power were observed. For a 5-m single-mode fiber, the system was tested with the following alterations:
- At interval 130–157, vibration was applied.
- At interval 262–310, external heat was applied.
- At interval 391–416, the fiber was bent.

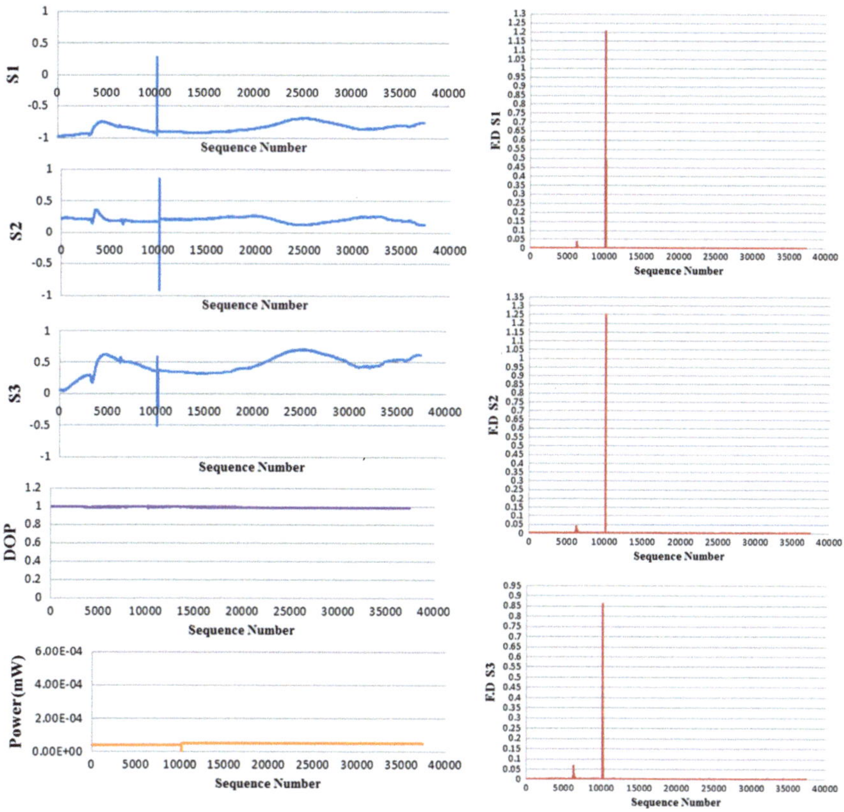

Fig. 10.5 Results of single-mode fiber with occasional alterations

The results of the 5-m single-mode fiber, displayed in Fig. 10.6, were consistent with the results obtained in the previous case. The intrusion system was robust against changing environmental conditions. The Stokes parameters experienced an abrupt change when the fiber was bent. The magnitude of the first-differences for all Stokes parameters far exceeded the sensitivity threshold. Again, only negligible changes in the degree of polarization and power were observed.

The same procedure that was applied for the single-mode fiber was repeated for a 5-m multimode fiber (Fig. 10.7). The system was tested with the following alterations:

- At interval 130–155, external vibration was applied to the fiber. S_1, S_2, and S_3 slightly changed and their first-differences remained below the threshold. Degree of polarization and power experienced negligible change.

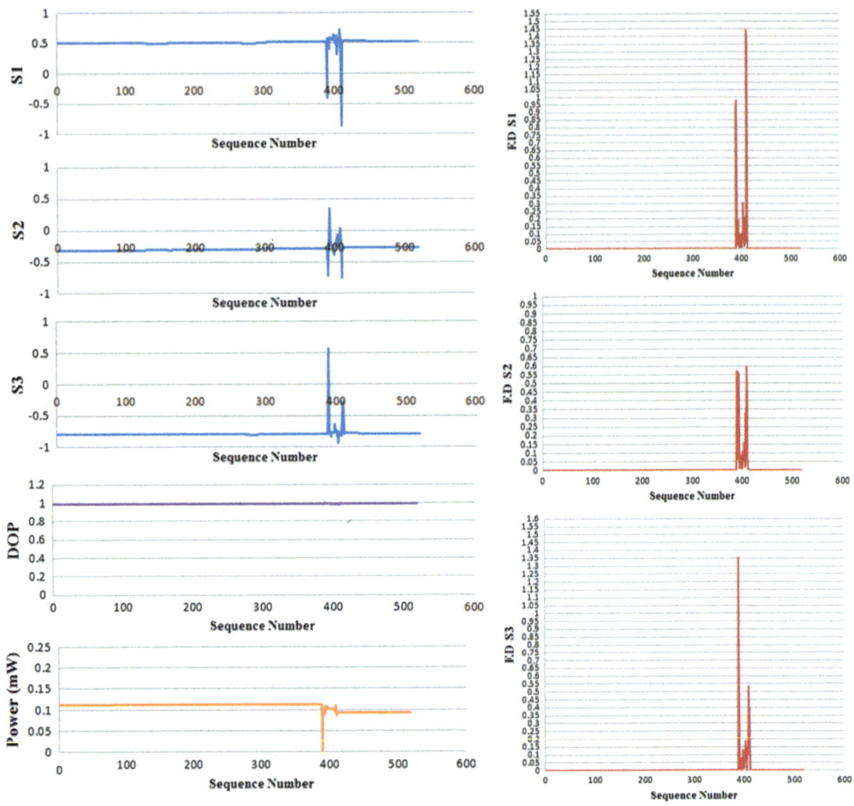

Fig. 10.6 Results of perturbed single-mode fiber

- At interval 260–285, the fiber was subjected to applied heat. This resulted in a change in the values of S_1, S_2, and S_3. Although the first-differences of some measurements in this interval exceeded the sensitivity threshold, not all three first-differences exceeded the sensitivity threshold simultaneously. These two events suggest that the intrusion system is robust against environmental conditions when using multimode fiber.
- At interval 390–415, the fiber was bent. All three Stokes parameters experienced an abrupt change. The magnitude of the first-differences of the Stokes parameters far exceeded the sensitivity threshold, suggesting a different generating process. Similar to the case of a single-mode fiber, only negligible changes in the degree of polarization and power were observed.

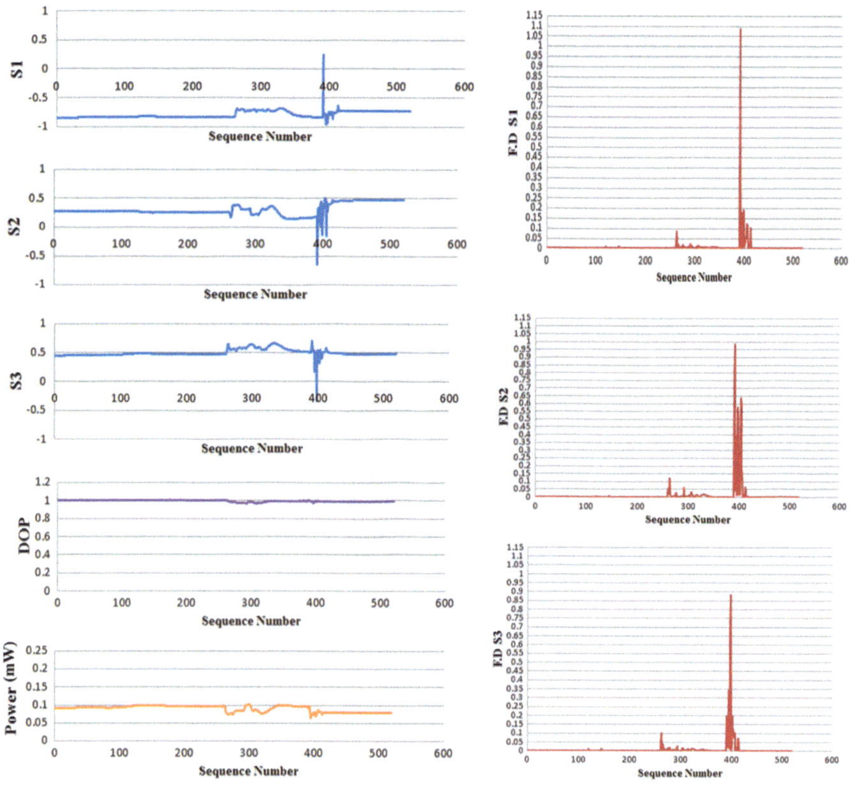

Fig. 10.7 Results of perturbed multimode fiber

10.6 Real-Life Applications of the Intrusion Detection System

The polarization-based intrusion detection system was implemented as shown in Fig. 10.8. A schematic diagram of the system is depicted in Fig. 10.9. An IP camera was used to transmit real-time video. The output of the IP camera was connected to a Cisco SGE2010P switch #1 in Fig. 10.8a. The intrusion detection system in Fig. 10.8b was applied on the link that connects switch #1 to Cisco SGE2010P switch #2. The corresponding port of switch #2 was connected to a computer for live viewing of the real-time video streamed by the IP camera in Fig. 10.8d. The information between the two switches was routed through an optical fiber link similar to the one deployed in current networks Fig. 10.8c. The passage of information through this optical fiber link was monitored using the intrusion detection system.

Information corresponding to the real-time video was sent through port 13 of the Fiber Patch Panel at the input of the armored cable. The information passed through

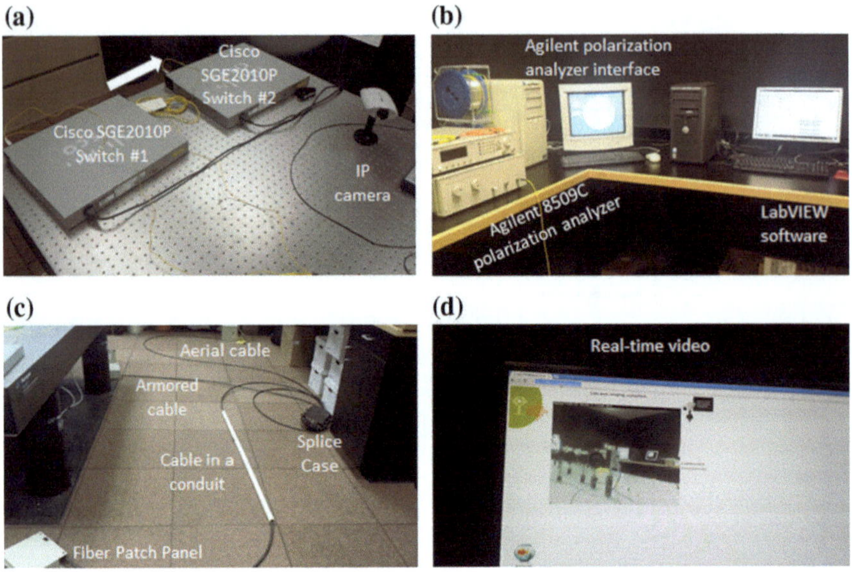

Fig. 10.8 Real-life application of intrusion detection system: **a** switches and IP camera layout; **b** intrusion detection system; **c** optical fiber link layout; and **d** real-time video

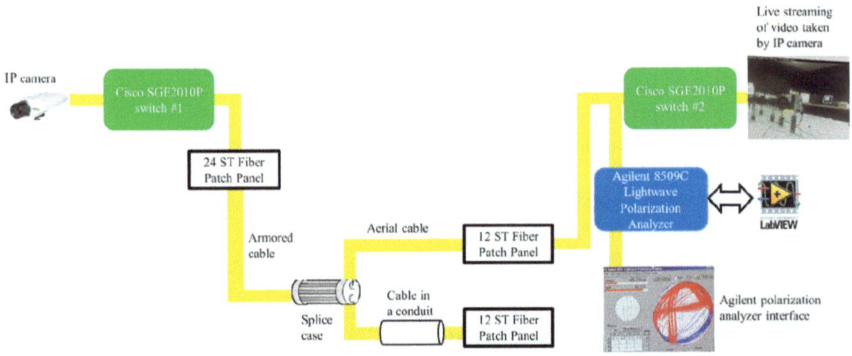

Fig. 10.9 Schematic diagram of real-life application of intrusion detection system

a Splice Case to either the aerial cable or to the cable in a conduit depending on the port number used at the input panel. (Ports 1–12 in the input panel connect the information to the conduit cable panel. Ports 13–24 in the input panel connect the information to the aerial cable panel.) The corresponding output port in the aerial cable panel was connected to the polarization analyzer of the intrusion detection system using a single-mode fiber. The results of the intrusion detection system as applied in the aforementioned network were consistent with previously described findings. When perturbations were applied to the optical fiber link, the system issued an intrusion alert to indicate that an intruder was trying to tap the link. The transmitted video captured by the IP camera showed no apparent difference for unperturbed and perturbed fiber links. This application of the intrusion detection system in this configuration showed that the system is able to detect intrusion activity on the optical fiber link even when no hint of such activity is apparent to an observer of the video output.

10.7 Summary

The intrusion detection system implemented and analyzed in this chapter prevents data theft in optical fibers. The behavior of certain attributes of light has been examined both in unperturbed and perturbed optical fibers of various lengths and types under different environmental conditions and changes in fiber geometry. The system has also been tested in a real-life scenario to monitor the link through which a real-time video was sent. The tests have proven that the state of polarization of light, represented by the Stokes parameters, is a better indicator of intrusion than the power and the degree of polarization. Furthermore, the first-differences of the Stokes parameters indicate the presence of an intruder when their values exceed a sensitivity threshold. In the case of intrusion detection, the system will issue an alert to stop the communication due to the possibility of data theft. Since optical fibers are the primary means of transmitting information, the proposed approach presents an innovative and cost-effective means to prevent data theft in contemporary telecommunication systems.

References

1. El Hajj, R., MacDonald, G., Verma, P., & Huck, R. (2015). Implementing and testing a fiber-optic polarization-based intrusion detection system. *Optical Engineering, 54*(9), 096107.
2. MacDonald. G. G., & Sluss, J. J., Jr. (2011). Method for polarization-based intrusion monitoring in fiberoptic links. US Patent No. US 7903977 B2, March 8, 2011.
3. Bhosale, S., Darunkar, B., Punekar, N., MacDonald, G., & Verma, P. (2016). Polarization based secure AES key transmission over optical fiber. In *2016 IEEE International Conference on Communications (ICC)* (pp. 1–6).
4. Hopkins, B., & Skellam, J. G. (1954). A new method for determining the type of distribution of plant individuals. *Annals of Botany, 18*(2), 213–227.
5. Galambos, J. (1994). Extreme value theory for applications. In *Extreme value theory and applications* (pp. 1–14). Berlin: Springer.

Chapter 11
Secure Key Transfer Over the Polarization Channel

Preceding chapters of the book have explored using the polarization channel of an optical fiber for transmitting secure information without the need for encryption keys. In addition, Chap. 10 has offered a means to use the polarization channel for detecting any malicious attempt to pilfer information in transit over an optical fiber. This chapter introduces the use of the polarization channel of an optical fiber to transfer data; more specifically, exchange symmetric keys between the two ends of the optical fiber. Use of the symmetric keys will allow any conventional symmetric encryption to take place between any number of data channels supported by the optical fiber. We do note that any encryption based on symmetric keys can be only computationally secure but, since the keys can be exchanged at a rapid rate, we can still achieve a high level of security on the data channel. Parts of this chapter have been previously published in [7, 8].

11.1 Symmetric Key Encryption

Chapter 10 has discussed use of the polarization channel to detect any attempt at violating the integrity of the polarization channel to ensure secure transfer of data over any of the wavelength channels of a Dense Wave Division Multiplexed System established over the fiber. As we have noted, the polarization channel detects any perturbation likely associated with an attempt to pilfer information from the wavelength channels. Absent any violation of the integrity of the polarization channel, the data channels are deemed to transfer information securely. If one could transfer data over the polarization channel itself, that data would be secure as well in the sense that any attempt to pilfer data will be detected in real time, and the data

P. K. Verma et al., *Multi-photon Quantum Secure Communication*,
Signals and Communication Technology,
https://doi.org/10.1007/978-981-10-8618-2_11

transfer stopped or re-routed over another facility. The polarization channel, however, is a low bandwidth channel and several constraints would prevent the transfer of data at a high enough rate to be of interest for carrying payload information. Secure transfer of symmetric keys over the polarization channel is, therefore, a viable alternative, allowing encryption of data which can then be transferred at high speed over a wavelength channel.

The QKD systems, such as BB84 discussed in Chap. 2, transfer secure keys as well, relegating the task of transferring secure payload information to symmetric encryption systems, such as the Advanced Encryption System (AES). The fundamental difference between QKD systems and the system discussed here is that, in QKD systems, the transfer of keys takes place over a quantum channel. Detection of intrusion on the quantum channel is enabled by monitoring the error rate on the quantum channel. In the system proposed here, keys are transported over the polarization channel and detection of any attempted intrusion is carried out by monitoring the finger print of the polarization channel.

This chapter discusses how one can exploit the polarization channel for carrying data. Even though the low bandwidth of the polarization channel will not be sufficient for customer data transfer, it can be used for transferring symmetric keys at a high enough data rate. The keys can be used to secure the data over the wavelength channels of the optical fiber using, for example, a symmetric encryption technology. This section briefly reviews symmetric key encryption, more specifically, the Advanced Encryption System for use in exchanging symmetric keys between the two end points of an optical fiber.

Symmetric Encryption technology, which is the technology of choice for most contemporary commercially used encryption systems, requires secure distribution of keys between a sender and a receiver. The only known means of unconditionally secure transfer of keys is based on Quantum Key Distribution (QKD) technology implemented in a limited way in a few commercially available devices such as the Clavis QKD system supplied by ID Quantique based in Geneva, Switzerland [1]. This technology is based on quantum mechanics requiring the generation, manipulation and detection of single photons. Since manipulation of single photons requires precision instruments that can be easily affected by ambient conditions that exist in a commercial telecommunication environment, contemporary QKD devices have found only niche applications to date. Additionally, use of single photons severely limits the distance and speed associated with QKD to just about a few hundred kilometers and speeds in the kilobits per second range.

This chapter addresses the secure transfer of AES keys over optical fibers using polarization as the channel of communication for transferring keys between the sender and the receiver. First, using the technology discussed in Chap. 10, the integrity of the polarization channel against possible leakage of information to an intending intruder is established. Subsequently, using the technology discussed in this chapter, the polarization channel is used to transfer keys between the two parties using states of polarization as the carrier of information. Since states of polarization or SoPs are known to vary in an unpredictable manner, a correction matrix is applied to the received SoPs to recover the original SoPs. The correction matrix is obtained from known SoPs exchanged between the two parties. We do note that it's entirely possible to use one of the wavelength channels of an optical fiber to transfer keys.

11.2 The Advanced Encryption System

The Advanced Encryption Standard is a symmetric encryption algorithm where the same key is used for encryption and decryption. AES encryption is based on combinations of substitution and permutation. AES has fixed block size of 128 bits, and key sizes of 128, 192, or 256 bits. The experimental system, deployed in our exploration, uses 128-bit data and 128-bit key size.

At the transmitting end, data to be transmitted is encrypted with a key and the generated ciphertext is transmitted over a wavelength channel. The key itself can be transmitted over another wavelength channel of the optical fiber or over the polarization channel. These two arrangements are shown in Figs. 11.1 and 11.2, respectively. The operation is as follows.

- Alice wants to send data X (128 bits) to Bob that is encoded with key Y (128 bits).
- Alice will run AES encryption algorithm with X (data) and Y (key) as inputs and will generate encrypted message called the Ciphertext C.
- This ciphertext will be transmitted over the fiber-optic communication channel to Bob. The key will be transmitted using another wavelength channel or the polarization channel.
- At the receiver, the received ciphertext is decrypted with the received key to get the transmitted data back.

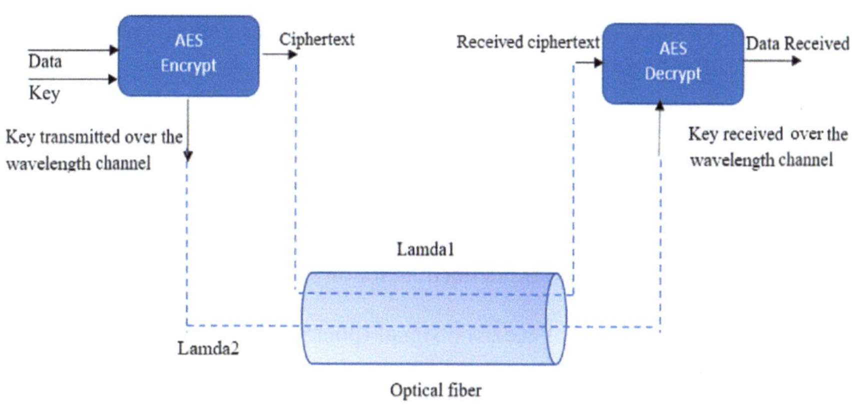

Fig. 11.1 Secure AES key transfer using a wavelength channel

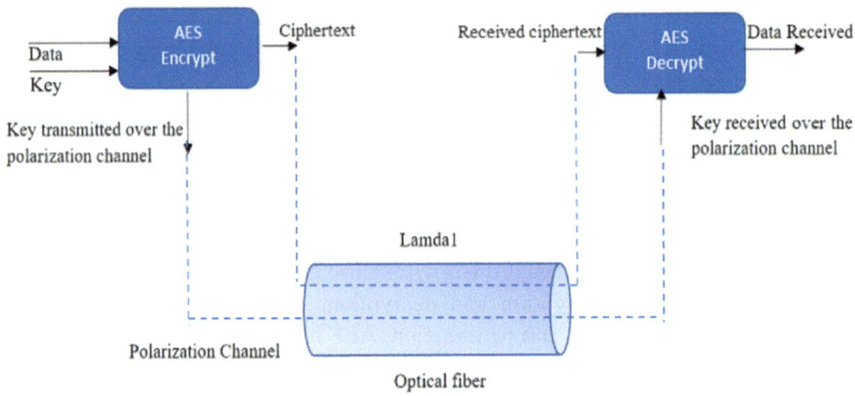

Fig. 11.2 Secure AES key transfer using the polarization channel

11.3 A Review of the Polarization Properties of Light

Chapter 2 has discussed the polarization properties of light. We review the polarization properties again in this chapter to the extent necessary for understanding the transmission of information over the polarization channel.

Polarization is an intrinsic property of light. The state of polarization (SoP) of electromagnetic waves can be thought of as the path traced out by the "tip" of the electric field vector as it propagates through some medium. The SoP of electromagnetic waves can be described by set of values called Stokes parameters. Stokes parameters were defined by George Stokes as "a mathematically convenient alternative to more common description of incoherent or partially polarized radiation in terms of its total intensity (I), (fractional) degree of polarization (P) and the shape parameters of the polarization ellipse" [2]. These parameters are determined by a set of intensity measurements ($S_0, S_1, S_2, and\ S_3$) acquired when light is passed through various types of polarizers. Collectively, these parameters are sometimes referred to as the Stokes vector,

$$S = \begin{bmatrix} S_0 \\ S_1 \\ S_2 \\ S_3 \end{bmatrix}$$

In terms of intensity, they are defined as follows:

$$S_0 = \text{Total power of received signal (polarized + unpolarized)}$$

$$S_1 = \text{Power received through a linear horizontal polarizer (LHP)} \\ - \text{Power received through a linear vertical polarizer (LVP)}$$

$$S_2 = \text{Power received through a } +45^\circ \text{ linear polarizer} \\ - \text{Power received through a } -45^\circ \text{ linear polarizer}$$

$$S_3 = \text{Power received through a right hand circular polarizer (RCP)} \\ - \text{Power received through a left hand circular polarizer (LCP)}$$

The Poincaré Sphere serves as a convenient method of mapping the Stokes polarization parameters and visualizing all possible SoPs. Briefly, each point on the Poincaré Sphere represents a unique SoP. Points along the equator represent linear SoPs, the poles of the sphere represent circularly polarized light, while points between the equator and poles represent elliptically polarized light. Fully polarized light is represented by points on the sphere and partially polarized light is represented by points interior to the sphere. When mapped onto the Poincaré Sphere, the Stokes parameters are represented as following:

$$S_0 = \text{Total Power (Polarized + unPolarized)}$$

$$S_1 = S_0 \cos(2\gamma)\cos(2\beta)$$

$$S_2 = S_0 \cos(2\gamma)\sin(2\beta)$$

$$S_3 = S_0 \sin(2\gamma)$$

where, 2γ and 2β are the spherical coordinates as shown in Fig. 11.3.

A few common SoPs and their Stokes representations are given in Table 11.1.

The SoPs in a single-mode fiber are very sensitive to any perturbation that is not symmetric about the fiber's axis. The primary mechanism for the change in polarization state is birefringence, which is basically used to describe effects of refractive index asymmetry in optical fibers. This asymmetry causes the two orthogonal states of polarization launched in fiber to propagate at different velocities. As a result, the originally launched polarization state is transformed as it

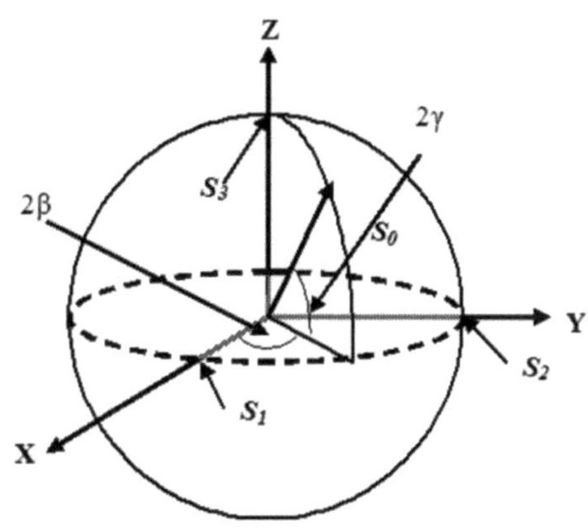

Fig. 11.3 Poincaré sphere

Table 11.1 Stokes vectors

LHP	LVP	RCP	LCP	Unpolarized
$\begin{bmatrix}1\\1\\0\\0\end{bmatrix}$	$\begin{bmatrix}1\\-1\\0\\0\end{bmatrix}$	$\begin{bmatrix}1\\0\\0\\1\end{bmatrix}$	$\begin{bmatrix}1\\0\\0\\-1\end{bmatrix}$	$\begin{bmatrix}1\\0\\0\\0\end{bmatrix}$

LHP Linear Horizontal Polarization
LVP Linear Vertical Polarization
RCP Right hand Circular Polarization
LCP Left hand Circular Polarization

propagates through fiber. The two primary causes of fiber birefringence are fiber geometry and stress. There is also a slow time-varying birefringence component that is random and unpredictable.

Many studies have been performed to characterize the magnitude and nature of evolving SoPs in optical fibers. Polarization fluctuations (in the form of Polarization Mode Dispersion or PMD) have been studied for aerial cables, buried cables and for spooled cables in an environmentally controlled setting [3]. Results show the variations in fiber temperature lead to PMD fluctuations. Aerial cables and long terrestrial links were more susceptible to PMD fluctuations. Studies performed on buried fiber found the polarization fluctuations to be slow, on the order of hours, with daily fluctuations generally between 2 and 10 degrees as measured on the Poincaré sphere [4]. The latter condition bears potential for use of polarization encoding techniques to transmit information.

11.4 Polarization Transfer Function and Fiber Characterization

Uncontrolled stresses or anisotropies are induced in the fiber during the manufacturing process and during deployment. These cause unwanted birefringence which is basically used to describe effects of refractive index asymmetry in optical fibers. Birefringence also depends upon many external factors such as temperature, vibration, and atmospheric pressure. There is also a slow time-varying birefringence component that is random and unpredictable. Due to this, optical fibers do not normally preserve the SoP of light propagating through a fiber. A standard light pulse is composed of two polarization modes traveling perpendicular to each other. In 'ideal' optical fibers, these polarization modes would travel at same speeds, however in case of 'real' world optical fiber, due to birefringence, these polarization modes travel at different speeds and causes PMD.

PMD fluctuations results in random spreading of optical pulses, limiting the rate of data transmission. This abnormality in optical fibers is significantly detrimental for extremely long fibers as the effects of PMD on SoP of optical beams are cumulative. The simplest way to reduce the PMD effect is by minimizing asymmetries in the index profile and stress profile of the fiber. However, even after steady improvements in manufacturing technology of fibers, faults remain. Therefore, there is a need to develop a technique of controlling polarization, making it possible to encode and send information using polarization encoding over an optical fiber. Any abrupt changes in SoP of light can be used as an indication of intrusion activity.

In [5], the authors have proposed a method of encoding information over the polarization channel. The proposed method includes a technique to compensate for unpredictable changes in SoPs due to birefringence mentioned above. In the following, we briefly discuss this compensation technique.

Evolution of SoPs in single mode fibers can be traced by investigating the motion of Stokes parameters on the Poincaré sphere as described by the Jones Calculus for fully polarized light or the Mueller Calculus for partially polarized light. We will use the Mueller formalism here. The Muller Calculus yields a 4×4 matrix that describe the polarization transfer function in terms of intensity. The elements of the Mueller matrix are real. This matrix provides the most general and complete description of the response of the medium as its elements contain polarization properties such as di-attenuation, retardance, and depolarization.

Suppose the SoP of launched light propagating through optical fiber be S and the received SoP be S'. The transfer function $M(t)$ can be described as:

$$S' = M(t) \cdot S$$

-Or-

$$S' = \begin{bmatrix} S'_0 \\ S'_1 \\ S'_2 \\ S'_3 \end{bmatrix} = M(t) \cdot S = \begin{bmatrix} m_{00} & m_{01} & m_{02} & m_{03} \\ m_{10} & m_{11} & m_{12} & m_{13} \\ m_{20} & m_{21} & m_{22} & m_{23} \\ m_{30} & m_{31} & m_{32} & m_{33} \end{bmatrix} \begin{bmatrix} S_0 \\ S_1 \\ S_2 \\ S_3 \end{bmatrix}$$

The Mueller matrix $M(t)$ describes the transfer function that is applied to the launched SoP after propagating through the fiber. Given knowledge of $M(t)$ and the received SoP $S(R)$, the launched SoP $S(L)$ can be recovered using:

$$S(L) = M^{-1}(t) \cdot S(R)$$

Hence it is possible to characterize the optical fiber before its use by deriving the appropriate Mueller matrix $M(t)$. The Mueller matrix of any optical element can be determined by measuring the resulting output for known input Stokes vectors. The process of deriving the Mueller matrix from these known vectors is referred to as fiber characterization.

Suppose six known SOPs such as linear horizontal polarization (LHP), linear vertical polarization (LVP), linear +45° polarization (L + 45°), linear −45° polarization (L − 45°), right hand circular polarization (RCP) and left hand circular polarization (LHP) are launched sequentially on a device under test (i.e. single mode fiber) and measured at the receiving end. Elements of the Muller matrix can be derived based on obtained results by following the procedure mentioned in [6]. Figures 11.4and 11.5 shows these six known SOPs launched into single mode fiber (SMF) which are respectively measured as output Stokes vector at the receiver end.

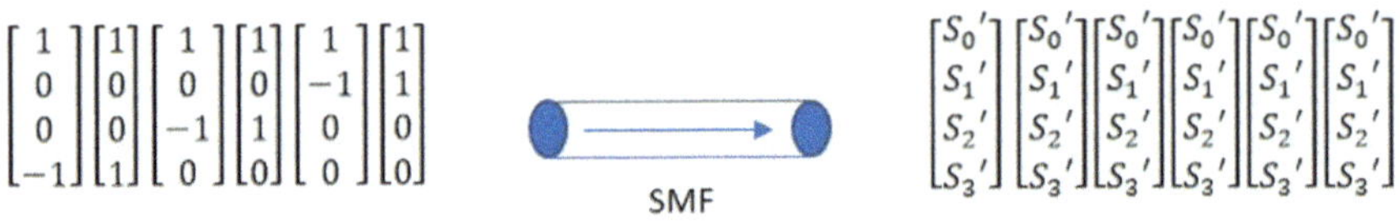

Fig. 11.4 Mueller matrix for SMF

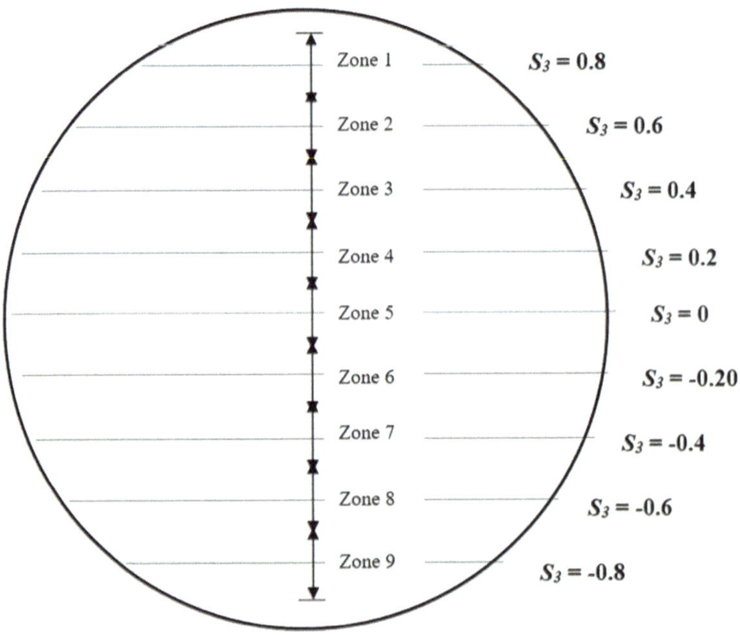

Fig. 11.5 256-POLSK [zone center values S_(3)]

The output stokes vectors of respective beams are represented as:

$$S_{\mathrm{LHP}} = \begin{bmatrix} S_0' \\ S_1' \\ S_2' \\ S_3' \end{bmatrix} = \begin{bmatrix} m_{00} & m_{01} & m_{02} & m_{03} \\ m_{10} & m_{11} & m_{12} & m_{13} \\ m_{20} & m_{21} & m_{22} & m_{23} \\ m_{30} & m_{31} & m_{32} & m_{33} \end{bmatrix} \begin{bmatrix} 1 \\ 1 \\ 0 \\ 0 \end{bmatrix} = \begin{bmatrix} m_{00}+m_{01} \\ m_{10}+m_{11} \\ m_{20}+m_{21} \\ m_{30}+m_{31} \end{bmatrix}$$

$$S_{\mathrm{LVP}} = \begin{bmatrix} S_0' \\ S_1' \\ S_2' \\ S_3' \end{bmatrix} = \begin{bmatrix} m_{00} & m_{01} & m_{02} & m_{03} \\ m_{10} & m_{11} & m_{12} & m_{13} \\ m_{20} & m_{21} & m_{22} & m_{23} \\ m_{30} & m_{31} & m_{32} & m_{33} \end{bmatrix} \begin{bmatrix} 1 \\ -1 \\ 0 \\ 0 \end{bmatrix} = \begin{bmatrix} m_{00}-m_{01} \\ m_{10}-m_{11} \\ m_{20}-m_{21} \\ m_{30}-m_{31} \end{bmatrix}$$

$$S_{\mathrm{L}+45^\circ} = \begin{bmatrix} S_0' \\ S_1' \\ S_2' \\ S_3' \end{bmatrix} = \begin{bmatrix} m_{00} & m_{01} & m_{02} & m_{03} \\ m_{10} & m_{11} & m_{12} & m_{13} \\ m_{20} & m_{21} & m_{22} & m_{23} \\ m_{30} & m_{31} & m_{32} & m_{33} \end{bmatrix} \begin{bmatrix} 1 \\ 0 \\ 1 \\ 0 \end{bmatrix} = \begin{bmatrix} m_{00}+m_{02} \\ m_{10}+m_{12} \\ m_{20}+m_{22} \\ m_{30}+m_{32} \end{bmatrix}$$

$$S_{\mathrm{L}-45^\circ} = \begin{bmatrix} S_0' \\ S_1' \\ S_2' \\ S_3' \end{bmatrix} = \begin{bmatrix} m_{00} & m_{01} & m_{02} & m_{03} \\ m_{10} & m_{11} & m_{12} & m_{13} \\ m_{20} & m_{21} & m_{22} & m_{23} \\ m_{30} & m_{31} & m_{32} & m_{33} \end{bmatrix} \begin{bmatrix} 1 \\ 0 \\ -1 \\ 0 \end{bmatrix} = \begin{bmatrix} m_{00}-m_{02} \\ m_{10}-m_{12} \\ m_{20}-m_{22} \\ m_{30}-m_{32} \end{bmatrix}$$

$$S_{\mathrm{RCP}} = \begin{bmatrix} S_0' \\ S_1' \\ S_2' \\ S_3' \end{bmatrix} = \begin{bmatrix} m_{00} & m_{01} & m_{02} & m_{03} \\ m_{10} & m_{11} & m_{12} & m_{13} \\ m_{20} & m_{21} & m_{22} & m_{23} \\ m_{30} & m_{31} & m_{32} & m_{33} \end{bmatrix} \begin{bmatrix} 1 \\ 0 \\ 0 \\ 1 \end{bmatrix} = \begin{bmatrix} m_{00}+m_{03} \\ m_{10}+m_{13} \\ m_{20}+m_{23} \\ m_{30}+m_{33} \end{bmatrix}$$

$$S_{\mathrm{LCP}} = \begin{bmatrix} S_0' \\ S_1' \\ S_2' \\ S_3' \end{bmatrix} = \begin{bmatrix} m_{00} & m_{01} & m_{02} & m_{03} \\ m_{10} & m_{11} & m_{12} & m_{13} \\ m_{20} & m_{21} & m_{22} & m_{23} \\ m_{30} & m_{31} & m_{32} & m_{33} \end{bmatrix} \begin{bmatrix} 1 \\ 0 \\ 0 \\ -1 \end{bmatrix} = \begin{bmatrix} m_{00}-m_{03} \\ m_{10}-m_{13} \\ m_{20}-m_{23} \\ m_{30}-m_{33} \end{bmatrix}$$

These six Stokes vectors contain all sixteen Mueller matrix elements. Now these vectors are passed through respective polarizers to measure their intensities. These intensities are then added and subtracted to get Muller Matrix elements. Let's consider LHP and LVP Stokes vectors passing through a linear horizontal polarizer and linear vertical polarizer, respectively.

Linear horizontal polarizer channel output

$$I_{1\,\mathrm{LHP}} = \frac{1}{2} \begin{bmatrix} 1 & 1 & 0 & 0 \\ 1 & 1 & 0 & 0 \\ 0 & 0 & 0 & 0 \\ 0 & 0 & 0 & 0 \end{bmatrix} \begin{bmatrix} m_{00}+m_{01} \\ m_{10}+m_{11} \\ m_{20}+m_{21} \\ m_{30}+m_{31} \end{bmatrix} = \frac{1}{2}(m_{00}+m_{01}+m_{10}+m_{11}) \begin{bmatrix} 1 \\ 1 \\ 0 \\ 0 \end{bmatrix}$$

$$I_{1\,\mathrm{LVP}} = \frac{1}{2}\begin{bmatrix} 1 & 1 & 0 & 0 \\ 1 & 1 & 0 & 0 \\ 0 & 0 & 0 & 0 \\ 0 & 0 & 0 & 0 \end{bmatrix}\begin{bmatrix} m_{00} - m_{01} \\ m_{10} - m_{11} \\ m_{20} - m_{21} \\ m_{30} - m_{31} \end{bmatrix} = \frac{1}{2}(m_{00} - m_{01} + m_{10} - m_{11})\begin{bmatrix} 1 \\ 1 \\ 0 \\ 0 \end{bmatrix}$$

Therefore, total intensities by adding and subtracting output Stokes vector from linear horizontal polarizer;

$$I_1^{(+)} = I_{1\,\mathrm{LHP}} + I_{1\,\mathrm{LVP}} = m_{00} + m_{10} \tag{11.1}$$

$$I_1^{(-)} = I_{1,\mathrm{LHP}} - I_{1\,\mathrm{LVP}} = m_{01} + m_{11} \tag{11.2}$$

Linear vertical polarizer channel output

$$I_{2\,\mathrm{LHP}} = \frac{1}{2}\begin{bmatrix} 1 & -1 & 0 & 0 \\ -1 & 1 & 0 & 0 \\ 0 & 0 & 0 & 0 \\ 0 & 0 & 0 & 0 \end{bmatrix}\begin{bmatrix} m_{00} + m_{01} \\ m_{10} + m_{11} \\ m_{20} + m_{21} \\ m_{30} + m_{31} \end{bmatrix} = \frac{1}{2}(m_{00} + m_{01} - m_{10} - m_{11})\begin{bmatrix} 1 \\ -1 \\ 0 \\ 0 \end{bmatrix}$$

$$I_{2\,\mathrm{LVP}} = \frac{1}{2}\begin{bmatrix} 1 & -1 & 0 & 0 \\ -1 & 1 & 0 & 0 \\ 0 & 0 & 0 & 0 \\ 0 & 0 & 0 & 0 \end{bmatrix}\begin{bmatrix} m_{00} - m_{01} \\ m_{10} - m_{11} \\ m_{20} - m_{21} \\ m_{30} - m_{31} \end{bmatrix} = \frac{1}{2}(m_{00} - m_{01} - m_{10} + m_{11})\begin{bmatrix} 1 \\ -1 \\ 0 \\ 0 \end{bmatrix}$$

Therefore, total intensities by adding and subtracting output Stokes vector from linear vertical polarizer can be obtained as follows;

$$I_2^{(+)} = I_{2\,\mathrm{LHP}} + I_{2\,\mathrm{LVP}} = m_{00} - m_{10} \tag{11.3}$$

$$I_2^{(-)} = I_{2\,\mathrm{LHP}} - I_{2\,\mathrm{LVP}} = m_{01} - m_{11} \tag{11.4}$$

These four Eqs. (11.1), (11.2), (11.3) and (11.4) are then solved for the matrix elements and we get,

$$m_{00} = \frac{1}{2}\left(I_1^{(+)} + I_2^{(+)}\right)$$

$$m_{01} = \frac{1}{2}\left(I_1^{(-)} + I_2^{(-)}\right)$$

$$m_{10} = \frac{1}{2}\left(I_1^{(+)} - I_2^{(+)}\right)$$

$$m_{11} = \frac{1}{2}\left(I_1^{(-)} - I_2^{(-)}\right)$$

Thus, we have determined matrix elements of the first quadrant of Mueller matrix. In a similar manner, other elements can be determined.

Once the Mueller matrix is defined, it represents the birefringence effect over optical fibers placed under the ambient environmental conditions which later can be used to negate the birefringence effects and get back the launched states of polarization.

In [5], the authors have proposed *256-Polarization shift keying technique* (256-POLS) which is an alternative to standard coherent modulation techniques such as amplitude shift keying (ASK), phase shift keying (PSK), frequency shift keying (FSK). This technique involves dividing the Poincaré sphere into a number of functional zones (8 operational or data zones and 1 calibration or control zone). The purpose of this system is to produce 256 different SoPs. The Stokes parameter S_3 acts as a reference point to divide the sphere. Each operational zone is assigned a particular number of segments (SoPs) depending upon circumference of the zone center. The calibration zone consists of two predefined polarization states which can be used to trigger the characterization procedure explained above, if necessary. Table 11.2 and Fig. 11.3 show detailed division of sphere. On the Poincaré sphere, each hexadecimal byte value's corresponding SOP can be defined by using the orientation angle ψ and ellipticity angle χ.

Table 11.2 Operational zones and symbols

Zone number	S_3 range	Polarized states	Hex byte values
1	$0.7 < S_3 < 0.9$	23	69–7F
2	$0.5 < S_3 < 0.7$	31	4A–68
3	$0.3 < S_3 < 0.5$	35	27–49
4	$0.1 < S_3 < 0.3$	38	01–26
5	$-0.1 < S_3 < 0.1$	2	00 and 80
6	$-0.3 < S_3 < -0.1$	38	81–A6
7	$-0.5 < S_3 < -0.3$	35	A7–C9
8	$-0.7 < S_3 < -0.5$	31	CA–EA
9	$-0.9 < S_3 < -0.7$	23	E8–FF

11.5 The System

This section presents the concept of the proposed application and the procedure of implementation.

Figure 11.6 shows application of the proposed technology in securing the key used for encrypting data using AES. The two communicating parties, Alice and Bob, are connected over a fiber-optic link. The transmission of keys uses the polarization channel without consuming any payload resources. Encrypted data is transmitted over the wavelength channel.

11.5.1 Method of Implementation

This sub-section describes the hardware and software involved in implementation of the system. This system is implemented and tested at OU-Tulsa Quantum Optics Lab. Figure 11.7 shows the various components involved:

System Components:

- Laser Source: Agilent 8509C light wave polarization analyzer having a laser source of wavelength 1550 nm is used for driving the polarization channel.
- Polarization Controller: Agilent 8169A polarization controller with single-mode fiber input and output, is used to change light from any polarized or un-polarized light source into any well-defined state of polarization. For this experiment, the controller was used to generate unique SoPs for every input character using the 256-POLSK technique.
- Polarization Analyzer: Agilent 8509C light wave polarization analyzer offers high-speed, calibrated polarization measurements of optical signals. For the experiment, this analyzer was used to measure the SoP represented by stokes parameter (S_1, S_2, and S_3), for deriving Mueller matrix.

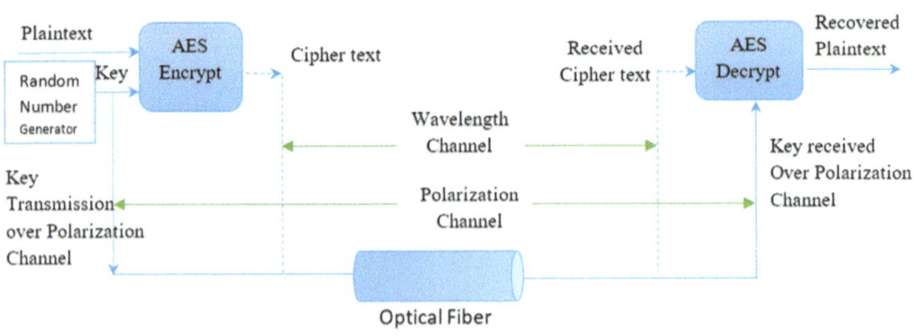

Fig. 11.6 Schematic of the implementation

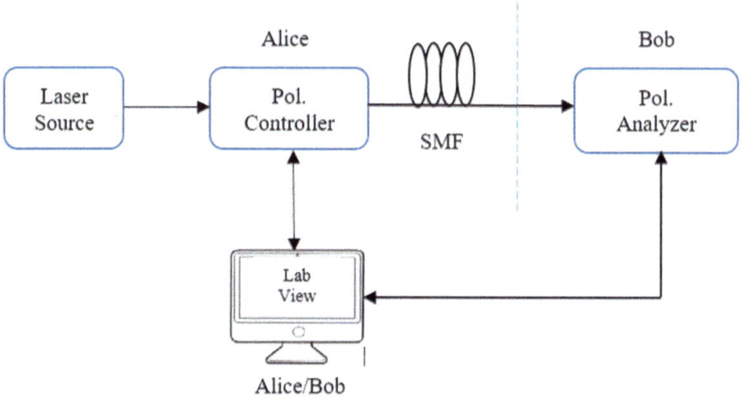

Fig. 11.7 Implementation system hardware and software

- Single Mode Optical Fiber (SMF): Single mode fiber of 3 m, 5, 10 and 50 km having FC/PC connector was used in the experiment. The operating wavelength is 1550 nm.
- LabVIEW Software: LabVIEW software version 10.0.1 installed on the PC was used to implement the procedure and control other devices using vendor supplied device drivers. PC connects polarization controller and polarization analyzer by means of GPIB bus. AES encryption and decryption algorithm was implemented over LabVIEW code.

Operation

Before establishing secure key transmission, it is necessary to authenticate the two communicating parties. This authentication procedure can be achieved over the classical communication channel. AES encryption decryption algorithm is implemented over LabVIEW software. Polarization monitoring intrusion detection setup is present at Bob's end to detect any kind of intrusion activity.

Detailed Procedure:

(1) *Authentication*

Alice first initiates communication request by authenticating Bob over wavelength channel.

(2) *Data and Key Generation*

Alice generates the key using random number generator. Alice chooses the plaintext.

(3) *AES Encryption*

Each character of the plaintext and the key is converted into binary through LabVIEW code. AES encryption algorithm is used to encrypt the plaintext with the key. The resulting ciphertext is transmitted to Bob over the wavelength channel.

(4) *Polarization based key transmission*

(a) First Alice and Bob characterize the polarization channel. Alice, using the LabVIEW program, first instructs the polarization controller to generate known SoPs. At the receive end, Bob uses polarization analyzer to measure the received SoPs and derive polarization reference frame (i.e., the Mueller matrix). This polarization reference frame captures the effects of birefringence over an optical fiber. The reference frame is used later to remove its effects.
(b) Once the reference frame is established, every character from the key is mapped over a 256-POLSK signal constellation points on Poincaré sphere. The LabVIEW program instruct the polarization controller to generate unique SoPs for each character and launch these SoPs over the optical fiber polarization channel for transmission.
(c) At the receive end, Bob measures received SoPs of each character using the polarization analyzer. Launched SoPs are recovered from the received SoPs using the following equation,

$$\text{Launched SOP} = \text{Received SOP} \cdot (\text{Mueller matrix})^{-1}$$

(d) Once Bob recovers the launched SoPs, LabVIEW performs reverse look up of the 256 POLSK mapping to get back the launched character.
(e) LabVIEW converts bit values to hex and displays the keys character one by one on PC. Bob receives a sequence of SoPs that are presumed to be the key followed by the checksum as, $(K'||CS')$.
(f) Bob declares that intrusion activity has been attempted and discards the key if,

$$CS' \neq \mathcal{F}[K']$$

(g) Bob declares that the received key is valid and no intrusion has taken place if,

$$CS' = \mathcal{F}[K']$$

(5) Intrusion Detection

The polarization channel is used to detect intrusion activity (potential placement of optical taps to tap a data channel) as well as for key transmission. Intrusion events effectively perturb the fiber geometry, significantly altering the SoPs and the transformation function. Figure 11.8 clearly shows that changes in SoPs are drastic in the case of fiber perturbation compared to changes in SoPs when a fiber is unperturbed and only environmental changes take place.

In case fiber perturbation takes place, Bob recognizes that the transfer function has been altered and the received key is discarded. Also, the LabVIEW program triggers an “intrusion detected” alarm. More specifically,

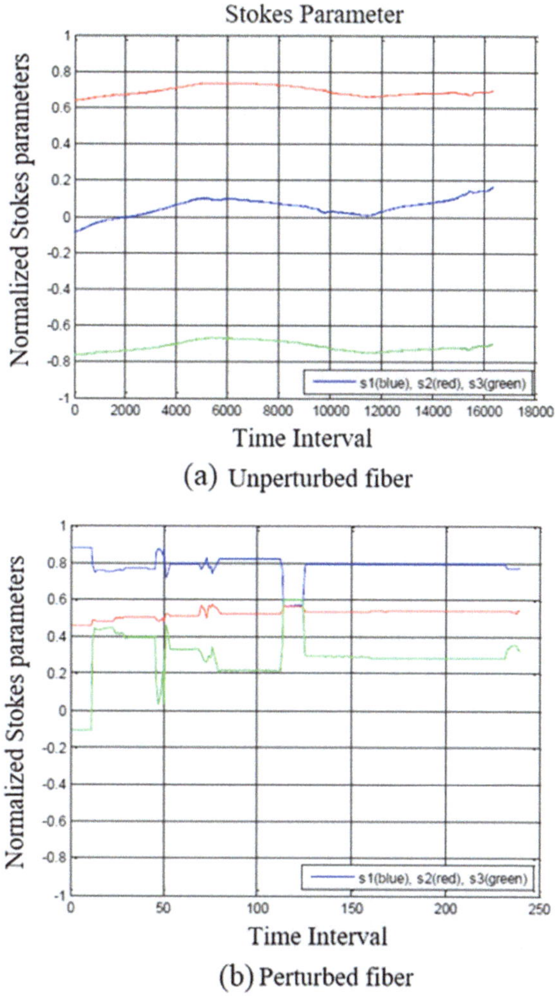

Fig. 11.8 Changes in SoPs **a** over unperturbed fiber **b** over perturbed fiber

- Let K be the key to be shared between Alice and Bob. In this method, Alice creates checksum character CS as a function of key K.

$$CS = \mathcal{F}[K]$$

Alice transmits the key K appended with the checksum CS as,

$$K||CS$$

(6) *Receiving data*

Bob will receive the cipher text over the wavelength channel. The valid key received over polarization channel is used to decrypt the cipher text. LabVIEW code is used for decrypting the cipher text.

11.6 Experimental Results

A photograph of the actual lab set up is shown in Fig. 11.9.

The secure key transfer technique was tested for multiple lengths of single mode fiber- 3 m, 5, 10, and 50 km. The results presented below are for AES key transfer carried over a 10-km fiber optic link. Results are plotted for both perturbed and unperturbed fiber.

Figure 11.10 illustrates the scenario where Alice wants to share key characters "POLARIZATIONKEYŽ" securely with Bob over the polarization channel. First, each character of the key is converted to a hexadecimal value and subsequently to bit values through the LabVIEW program. The corresponding SoP values (S_1, S_2, and S_3) for each character are sent to Bob. Bob, after receiving these SoPs, recovers the launched characters with procedure explained above.

Figure 11.11 shows results for unperturbed fiber where Bob received the AES key properly, i.e., "POLARIZATIONKEYŽ". The transmitted SoPs (S_1, S_2, S_3) and received SoPs (S_1, S_2, S_3) for each character are plotted against their values (Fig. 11.9). It can be observed that for the unperturbed fiber transmitted SoPs (S_1, S_2, S_3) are approximately similar to the received SoPs (S_1, S_2, S_3).

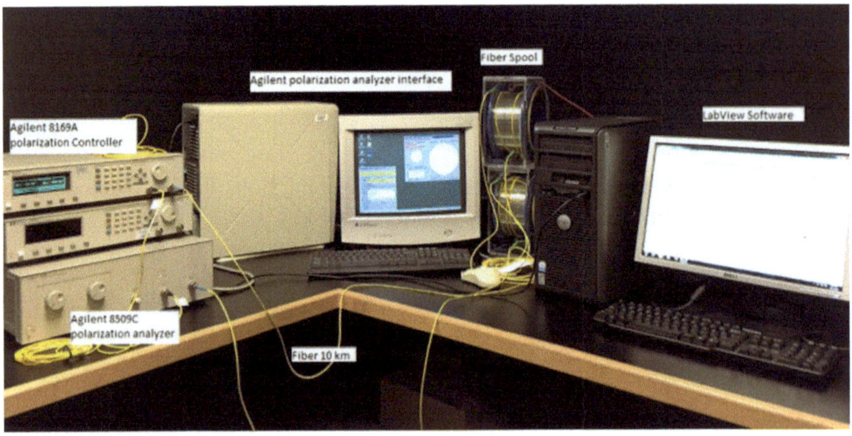

Fig. 11.9 Lab set up

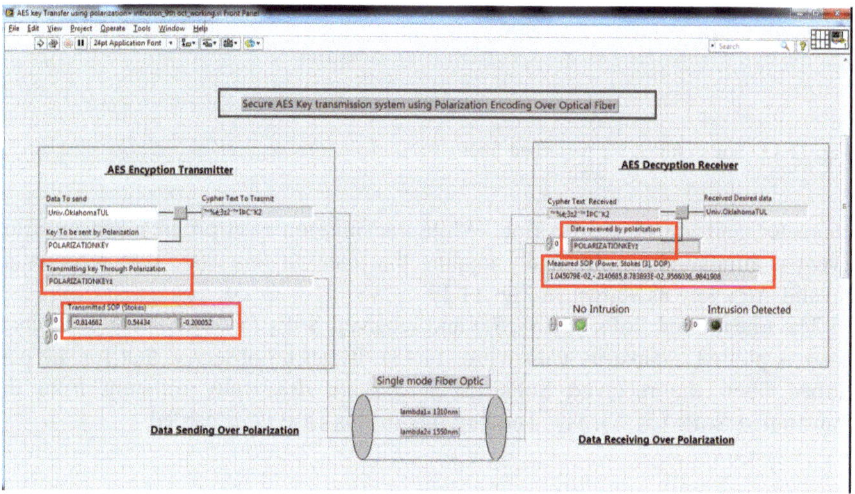

Fig. 11.10 Unperturbed fiber front panel LabVIEW

Fig. 11.11 Transmitted and received SoPs for unperturbed fiber

Figure 11.12 describes the scenario where Alice wants to share key characters "POLARIZATIONKEYm" securely with Bob over the polarization channel. The LabVIEW program calculates the corresponding SoP values (S_1, S_2, S_3) for each

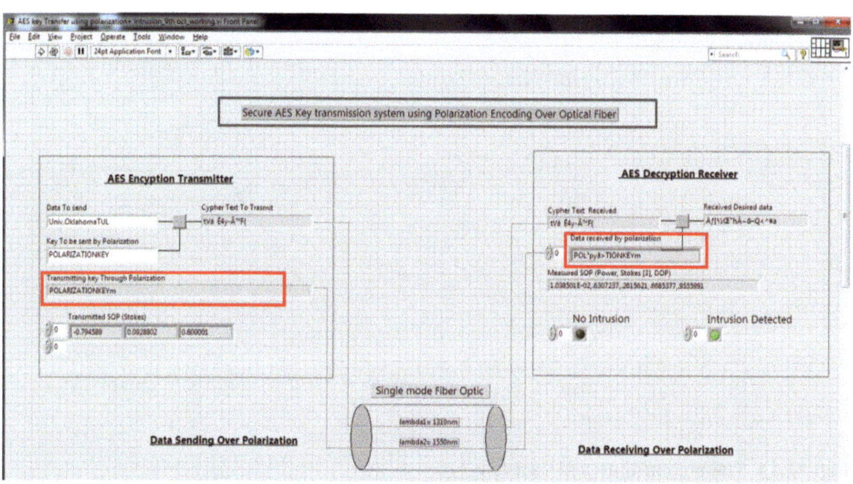

Fig. 11.12 Perturbed fiber front panel LabVIEW

Fig. 11.13 SoPs plotted for perturbed fiber

character, and transmits it to Bob. While transmission is in progress, an intrusion event is simulated by physically tapping the fiber. In this case, Bob receives an invalid AES key as shown in Fig. 11.13.

The transmitted SoPs (S_1, S_2, S_3) and received SoPs (S_1, S_2, S_3) for each character is plotted against its value (Fig. 11.13). It can be observed that for the perturbed fiber, the received SoPs (S_1, S_2, S_3) are drastically different from the transmitted SoPs (S_1, S_2, S_3). The key bits in this case is discarded.

11.7 Data Rate and Calibration Time

Calibration time and data rate for transmitting the key over the polarization channel is measured for 3 m, 5, 10 and 50 km single mode optical fiber. The results are presented in Fig. 11.14.

11.8 Summary

This chapter has offered the concept, design and implementation of secure AES key transmission system based on 256 POLSK polarization encoding. A successful laboratory implementation and results have shown the possibility of using the polarization channel to detect intrusion and transport AES keys over the polarization channel securely [7, 8].

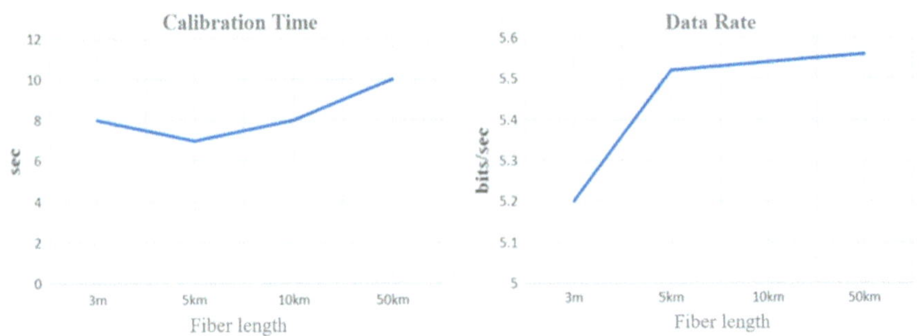

Fig. 11.14 Calibration time and data rate

References

1. http://www.idquantique.com/quantum-safe-crypto/qkd-overview/.
2. Chandrasekhar, S. (2013). *Radiative transfer*. Courier Corporation.
3. Cameron, J., et al. (1998). Time evolution of polarization mode dispersion in optical fibers. *Photonics Technology Letters, IEEE, 10*(9), 1265–1267.
4. Nicholson, G., & Temple, D. J. (1989). Polarization fluctuation measurements on installed single-mode optical fiber cables. *Lightwave Technology, Journal of, 7*(8), 1197–1200.
5. MacDonald, G. G., & Sluss, J. J., Jr. (2011, March 8). *Method for polarization-based intrusion monitoring in fiberoptic links*, U.S. Patent No. US 7,903,977 B2.
6. Collett, E. (2003). *Polarized light in fiber optics*. SPIE Press.
7. Bhosale, S., Darunkar, B. A., Punekar, N. V., Gregory, M. D., & Verma, P. K. (2016, May 23–27). Polarization based secure AES key transmission over optical fiber. In: *Presented at the IEEE ICC 2016 Commun. and Inf. Syst. Secur. Symposium and published in the Proceedings of IEEE ICC 2016*. Kuala Lumpur. Reprinted with permission.
8. Bhosale, S. (2015). *Secure AES key transmission using polarization encoding over optical fiber* (A thesis submitted to the graduate faculty). University of Oklahoma, 2015. Reprinted with permission.

Chapter 12
An Ultra-Secure Router-to-Router Key Exchange System

This chapter presents an ultra-secure router-to-router key exchange system. The key exchange process can be initiated by either router at will and can be carried out as often as required. The cryptographic strength of the proposed protocols lies in the use of multi-stage transmission where the number of variables exceeds the number of stages by one, ensuring that the number of possible measurements is one less that the number of variables. The proposed system carries out all processing in electronics and is not vulnerable to the man in the middle attack. The treatment presented in this chapter is based on the authors' work in [1, 2].

12.1 Introduction

With the rise of globalization, microelectronics, and the information age, the need for rapid, long-distance transmission of unconditionally secure information has never been greater. Whether dealing with military intelligence, corporate secrets shared between two (or more) company offices, remote control of vital national infrastructure components such as power and traffic control systems, or mechanical instructions transmitted to offsite medical devices for tele-surgery, device updates and health reports, there are many situations where the rapid, accurate, and secure transmission of information between two parties is a basic necessity. In extreme cases, alteration, or even decryption of this information by unauthorized parties may result in damages of billions of dollars and the lives of others.

As discussed in Chap. 1 of this book, historically, only two encryption schemes have been proposed which offer unconditional security, both unsuitable for practical telecommunications. The first, the one-time pad, proposed by Joseph Mauborgne [3], utilizes a single-use encryption key equal to the message length which both the sending and receiving parties may use to encrypt and decrypt the message. The disadvantages of this system in a long-term high-data rate communication system are obvious, with each message requiring a pre-shared key equal to the message

P. K. Verma et al., *Multi-photon Quantum Secure Communication*,
Signals and Communication Technology,
https://doi.org/10.1007/978-981-10-8618-2_12

length. The second, recently proposed unconditional cryptographic system is quantum cryptography, where security is achieved through the laws of quantum mechanics, which allow for very accurate determination of eavesdropping along a quantum channel, as well as the simultaneous determination of small shared and secure random values. Currently available quantum encryption protocols include BB84, proposed in 1984 by Bennett and Brassard [4], the variant SARG04 [5], and the later-developed B92 [6]. All three solutions, while theoretically secure, place severe requirements on the needed hardware and are generally considered to be unsuitable for commercial use. The limitations of these systems include, for example, reliance on single-photon generators (greatly limiting practical data rate) and, most importantly, the presence of a physical, well characterized quantum channel between endpoints, with a maximum practical distance of a few hundred kilometers. While research related to the use of quantum sources [7] and channel extension [8] continues, this technology remains in its early stages and is not ready for widespread commercial applications. We note that BB84, SARG04 and B92 are secure only to the extent that transmission of *random* information is concerned. The random information can function as a one-time pad for the transmission of unconditionally secure information. The limitation that the key can't be used more than once means that the length of the payload cannot be more than the length of the key itself. This puts an unrealistic constraint on the commercial exploitation of the technique, reducing the application of the technique to the transmission of keys which can then be used in a conventional symmetric or asymmetric communication system. Any such system cannot be deemed to be unconditionally secure.

While unconditional security may be an unachievable goal in the practical sense, communication security may be achieved at an arbitrarily high level via existing symmetric and asymmetric encryption systems, offering a mathematically provable level of security generally corresponding to the algorithm and key length used. Currently, the most widely used form of global network communication between two distant parties relies on public-key, asymmetric key cryptography such as RSA in the form of public-private certificate pairs issued through large trusted corporations. Noteworthy corporations offering this SSL certificates with Elliptic Curve Cryptography (ECC), RSA and DSA support include Symantec (formerly Verisign), GoDaddy, and Comodo.

Although presenting a viable and widely-used solution to secure communication, allowing for message encryption and authentication, the security certificate system requires the presence of a trusted third party for the distribution and management of keys. The compromise or loss of trust in such a third party, or the inability to contact the distribution network when needed, may result in a large-scale breakdown of reliable and secure communications [9]. Furthermore, the increasingly large RSA key length requirements of public certificates to guarantee secure communication may be a barrier to practical implementation on limited-resource devices.

This chapter presents an ultra-secure router-to-router key exchange system. The key exchange process can be initiated by either router at will and can be carried out as often as required. We compare the efficacy of the proposed approach with contemporary quantum key distribution (QKD) systems and show that a very high level of security is attainable without resorting to single photon generators and other attendant instrumentation associated with QKD. Furthermore, the proposed system addresses the extremely limited geographical reach of commercially available QKD systems and other environmental restrictions they must operate in. The proposed system carries out all processing in electronics and is not vulnerable to the man in the middle attack. The medium of transfer can, of course, be optical fibers, as is common in telecommunication systems.

12.2 Related Work

12.2.1 Discrete Logarithms

Discrete logarithms constitute a well-known technique in number theory [10, 11]. Discrete logarithms in Z_p are related to the primitive roots of the prime number p. All prime numbers have primitive roots. One characteristic of a primitive root α of a prime number p is that successive powers of α from 1 to $p - 1\,\mathrm{mod}(p)$ generate the numbers from 1 to $p - 1$, all of which are distinct, but in some unpredictable order. In other words, for any integer b and a primitive root α of a prime number p, there is a unique exponent i such that,

$$b = \alpha^i \mathrm{mod}(p), \quad 0 \leq i < p - 1 \tag{12.1}$$

The exponent i is referred to as the discrete logarithm of the number b for the base $\alpha\,\mathrm{mod}\,p$.

One of the properties of discrete logarithms is that, given α, i, and p, the computation of b is easy. However, given b and α, computation of i is difficult. This difficulty is of the same order of magnitude as factoring primes required for RSA [12]. We note that, unlike the Diffie-Hellman technique (discussed in the next subsection), these parameters are not global. They are known only to the concerned router pairs.

We derive our clue from the fact that if the information to be transmitted (in other words, the key) gets embedded in the exponent i (or the discrete logarithm) and is transmitted as the number b, deriving the key will be as hard as any contemporary encryption scheme. This combined with the fact that the key exchange can take place as frequently as desired with minimal overhead in communication, we posit that the proposed scheme has the potential of approaching the level of security of contemporary quantum key distribution (QKD) systems. The following section discusses one of the contemporary key management/distribution protocols.

12.2.2 Contemporary Key Distribution Protocols

In this section, the Diffie-Hellman key distribution [13, 14] technique is discussed. The Diffie-Hellman key distribution algorithm is based on using prime numbers as well as primitive roots in a similar context as in the proposed multi-stage protocol.

Figure 12.1 summarizes the Diffie-Hellman key exchange algorithm. In this scheme, there are two publicly known numbers (global elements); q is a prime number with its primitive root α. X_A and X_B are two random variables chosen by Alice and Bob, respectively. Using the key exchange system presented in Fig. 12.1, Alice and Bob will share a key $K = (Y_B)^{X_A}\text{mod}(q) = (Y_A)^{X_B}\text{mod}(q)$.

The Diffie-Hellman key exchange algorithm depicted in Fig. 12.1 is, however, vulnerable to the man in the middle attack. An attacker, Eve, will proceed as shown in Fig. 12.2 to share a key K_1 with Alice and another key K_2 with Bob. Thus, all future communication between Alice and Bob is compromised.

Shared keys between different parties in a network can also be established with the help of a trusted third party. This third party is called the key distribution center [11]. Furthermore, symmetric key distribution using asymmetric encryption can be used as a key distribution scheme as proposed in [15, 16]. It is worth noting that the scheme proposed in [15] is vulnerable to the man in the middle attack, whereas the scheme proposed in [16] assumes that Alice and Bob already exchanged a public key using a public key distribution technique. Several public key distribution techniques can be used, e.g., public announcement, publicly available directory, public-key authority, and public certificate.

The protocol proposed in this chapter overcomes the overhead of maintaining an updated list of the public keys of each node in the network as well as sharing them globally by the means mentioned earlier. We do note that the proposed system, however, is deployed for encrypting messages on a link which connects two routers. It does not offer an end-to-end application level encryption scheme.

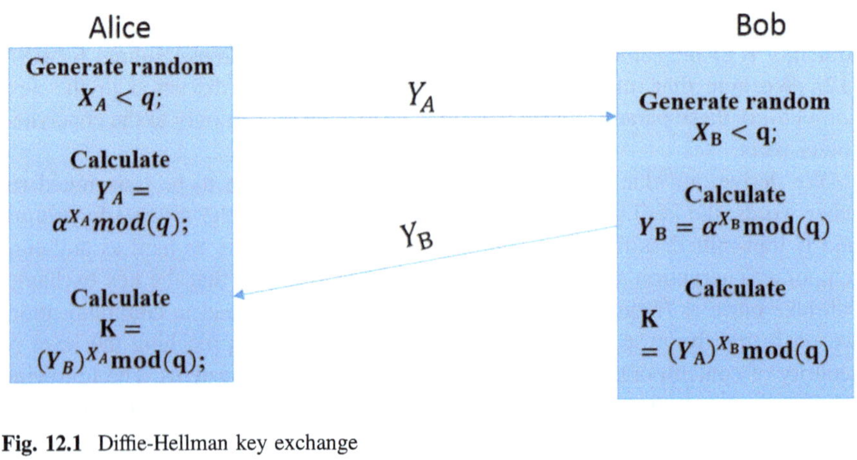

Fig. 12.1 Diffie-Hellman key exchange

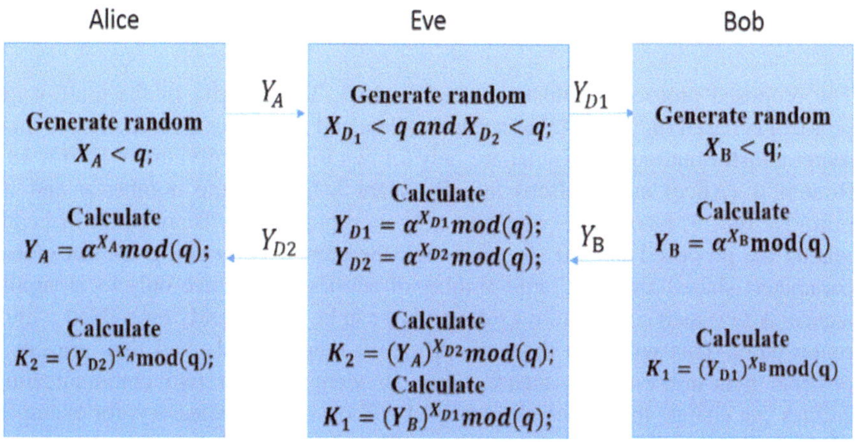

Fig. 12.2 Man in the middle attack in case of a Diffie-Hellman key exchange system

12.3 The Proposed Protocol

In this section, the proposed multi-stage router to router key distribution protocol is presented. The proposed protocol is a decentralized approach to key distribution. Sharing of public keys or symmetric keys throughout the network is not needed. Any two routers having a primitive root α and a prime number p in common can initiate the key exchange. A simulation of the proposed approach, discussed in the following subsection is presented in Table 12.1. The basis of the proposed key exchange is discussed next.

Table 12.1 Simulation of the key at Alice and Bob along with the value of each leg of the protocol

Cycle	Key at Alice's	i_1	i_2	L_1	L_2	L_3	Key at Bob's
1	46	0.958478	4.238835	89.08386	52.85938	46.83984	46
2	46	4.048701	7.071245	59.04688	58	57	46
3	46	1.851114	9.189888	75.58826	6	73	46
4	46	1.737055	7.436969	31.79602	32	53	46
5	46	2.544817	9.741039	27.98389	65	89	46
6	46	2.730982	10.93591	73.00684	66	91	46
7	46	0.983038	7.114608	37.65381	31	32	46
8	46	2.436789	5.763868	52.89453	65	71.375	46
9	46	3.325503	10.80495	92.74902	45	9	46
10	46	2.784334	5.076208	61.81592	81	27.75	46

12.3.1 Multi-stage Protocol

The proposed protocol is illustrated in Fig. 12.3. The security of the multi-stage protocol is based on the fact that a sender Alice and a receiver Bob each have their secret transformations (keys) that are only known to them, individually [7, 17, 18]. Routers 1 (Alice) and 2 (Bob) initially share a large prime number p and its primitive root α. Knowledge of p and α by a cryptanalyst will not result in invalidating the proposed scheme. In any event, the parameters α and p are not global parameters shared among a large number of entities. They have only local significance. A few such combinations (of α and p) can be permanently embedded within router pairs. This will help them identify each other initially and thus prevent a possible man-in-the-middle attack. Once so identified, the two communicating entities can change the values of α and p at will or, if at all necessary, for example, after a major attack and/or system outage. The proposed scheme of key exchange now follows.

$$K = \alpha^x \bmod p \tag{12.2}$$

The key exchange initiating party, say Alice, chooses a random positive number x, where x is less than p. Alice then generates $\alpha^x \bmod p$ as the key to be shared. In other words, the key K can be represented as,

$$N = \alpha^{i_1} \bmod p \tag{12.3}$$

Simultaneously, Alice chooses another random number $i_1 < p$ and generates an intermediate number N as,

$$N^{-1} = \alpha^{-i_1} = \alpha^{p-1-i_1} \bmod p \tag{12.4}$$

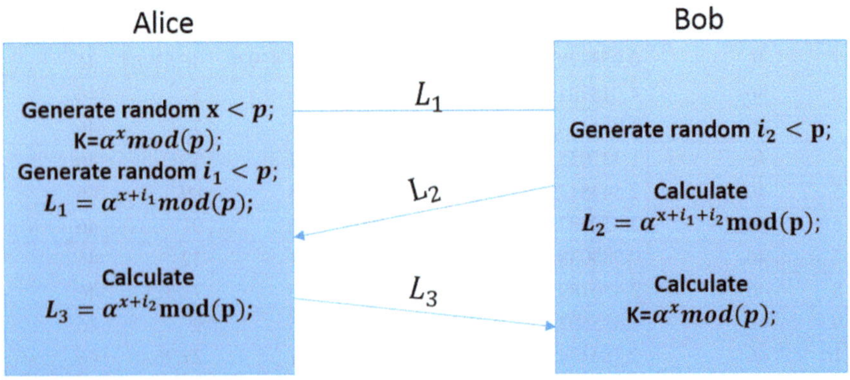

Fig. 12.3 Key exchange scheme using discrete logarithms

Alice also computes and stores N^{-1}, or the inverse of $N \bmod p$. While computation of the inverse of $N \bmod p$ might appear to be a formidable task, in practice, it isn't due to the following relationships.

The latter relationship follows because for any prime number p and its primitive root α, we have,

$$\alpha^{p-1} = 1 \bmod p \tag{12.5}$$

Equation (12.3) converts the inversion process to a simple exponentiation process. Alice preserves the key K and the intermediate numbers $N = \alpha^{i_1} \bmod p$ and $N^{-1} \bmod p$ in a table. On the first leg of transmission, Alice generates and transmits $L_1 = \alpha^{x+i_1} \bmod p$. This and the subsequent transmissions are shown in Fig. 12.3.

Bob has similarly chosen a random number $i_2 < p$ and generated the corresponding numbers $\alpha^{i_2} \bmod p$ and generated its inverse $\alpha^{-i_2} \bmod p$ following the procedure outlined for Alice. On the second leg from Bob to Alice, Bob generates and transmits, $L_2 = \alpha^{x+i_1+i_2} \bmod p$ as shown in Fig. 12.3.

Upon receiving L_2, Alice multiplies it by N^{-1} and getting $\alpha^{-i_1} \bmod p$, thus creating,

$$L_3 = \alpha^{x+i_1+i_2}\alpha^{-i_1} \bmod p = \alpha^{x+i_2} \bmod p \tag{12.6}$$

which she then transmits to Bob on the third leg. Upon receiving L_3, Bob can easily evaluate the key K as, $K = L_3\alpha^{-i_2} = \alpha^x \bmod (p)$, which is the key Alice intended to share with Bob. Bob can similarly transmit a key to Alice.

A simulation of the values of L_1, L_2, and L_3 using Matlab has been done. The results for random values of i_1 and i_2, $\alpha = 5$, $x = 15$, and $p = 97$ are shown in Table 12.1.

12.3.2 Man in the Middle Attack on Multi-stage Protocols

In this section, the cryptographic strength of the proposed protocol is discussed by considering its vulnerability to the man in the middle attack. It can be noted from the previous section that the key to be transmitted is based on a random number x. This random number is never transmitted into the open nor is the actual key, which is $\alpha^x \bmod p$. The transmission on the first leg L_1 is $\alpha^{x+i_1} \bmod p$, which is a logarithmic function of the key. An intruder having access to L_1 (see Fig. 12.3) is handicapped because i_1 is an unknown quantity and even if the intruder had access to i_1, the intruder will still have to compute the discrete logarithm. The same level of difficulty is associated with the intruder's access to any of the three legs of transmission in isolation.

The case of the intruder having access to each of the three legs of transmission at the same time is considered; in other words, the intruder can access the L_1, L_2 and

L_3 messages that belong to the same information transmission (Fig. 12.4). It is apparent from the three equations that, after some algebraic manipulation, capture of each of these three streams will result in the intruder Eve capturing $\alpha^x \bmod p$ which is the intended key. An intruder can make use of this access by either sharing the key $\alpha^x \bmod p$ with Alice, then sharing it with Bob. Or, the intruder can effect a man in the middle attack where he/she shares a key with Alice (K_{A-E}) and another key with Bob (K_{B-E}). This case is shown in Fig. 12.4, and a simulation of the results L_{A-E1}, L_{A-E2}, and L_{A-E3} (transmissions between Alice and Eve), L_{B-E1}, L_{B-E2}, and L_{B-E3}(transmissions between Bob and Eve), using Matlab has been done. The results for random values of i_1, i_2, i_{E1}, and i_{E2} with $\alpha = 5, x = 15$, $x_{E2} = 20$, and $p = 97$ have been presented. As seen from Table 11.2, a secret key will be shared between Alice and Eve and a different secret key will be shared between Eve and Bob. During the communication Alice and Bob will not be able to detect the presence of Eve in the middle.

These vulnerabilities are effectively addressed in the proposed protocol in the next section.

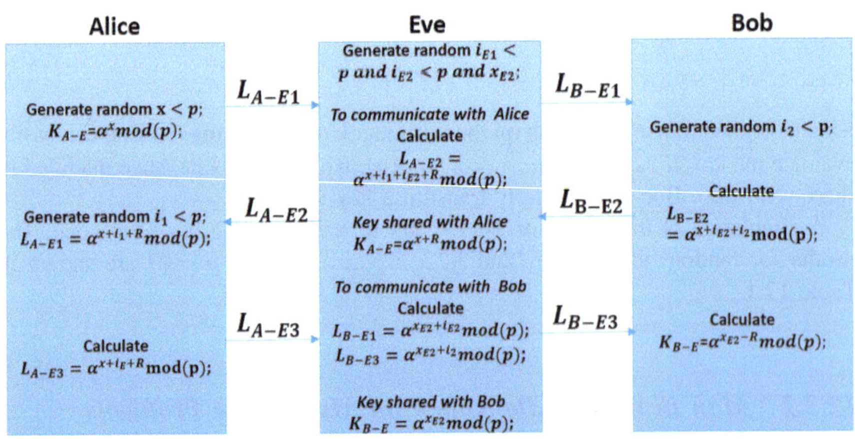

Fig. 12.4 Man in the middle attack on the proposed system

Table 12.2 Simulation of a successful man in the middle attack in case of a multi-stage protocol without an initialization vector

Cycle	Key shared between Alice and Eve	L_{A1}	L_{A2}	L_{A3}	L_{B1}	L_{B2}	L_{B3}	Key shared between Eve and Bob
1	46	40.56	44.5	79.70	18	69	50	93
2	46	8.32	39	30	60	34	72	93
3	46	68.18	65	56	29	38	42	93
4	46	40.31	37	33.62	87	96	44	93
5	46	23.75	87.12	10.65	5	88	40	93

12.4 Proposed Protocol Using an Initialization Vector and Its Cryptographic Strength

One can think of numerous ways to deny Eve from simultaneously accessing all of the three legs of the transmission. An obvious way is to send the information on each of the three legs on separate fibers or separate DWDM channels of one fiber. Furthermore, such assignments can be randomized.

Another way to frustrate Eve from accessing the key even after a successful capture of the three messages is to retain a subset of the last interchange of the key between Alice and Bob, and couple it with the new key $\alpha^x \bmod p$ to be transmitted with the following transformation, $\alpha^x \alpha^R \bmod p$, where α^R represents a remnant of the key successfully transferred during the last interchange. Since only Bob has access to R, he can easily derive the key $\alpha^x \bmod p$. In addition to providing security, use of the remnant R will also obviate any man-in-the-middle attack. It should be noted that since R is changing periodically, it is substantively different from the Initialization Vector used in conventional cryptography where it is fixed and therefore subject to a range of cryptanalytic attacks. The proposed approach is discussed in detail in the following section.

12.4.1 *Description*

As indicated in Sect. 12.3, the problem with the multi-stage algorithm is that if an intruder is present on all the stages of the communication at a given time, he/she will be able to get the ciphertext at each stage of the protocol [1]. Knowing the cipher text at each stage of the protocol, the intruder will be able to solve the following system of simultaneous equations:

$$L_1 = \alpha^{x+i_1} \bmod p = m \tag{12.7}$$

$$L_2 = \alpha^{x+i_1+i_2} \bmod p = m' \tag{12.8}$$

$$L_3 = \alpha^{x+i_2} \bmod p = m'' \tag{12.9}$$

where m, m', and m'' are the values measured at L_1, L_2, and L_3 respectively. An underlying assumption, of course, is that α and p are known.

Since these are not global parameters, in the absence of inside information, Eve will be locked out from getting the key. In the following we address the situation when these parameters are available to the intruder and show how the use of the initialization vector will still prevent Eve from accessing the key.

In the previous section, we proposed a way to prevent such type of attacks by using a remnant from a previously used key α^R. Thus the number of stages of the protocol are only three, while the number of random variables used to protect the message are four. On the first leg of transmission, Alice generates and transmits $L_1 = \alpha^{Ri_1} \bmod p$. On the second leg from Bob to Alice, Bob generates and transmits, $L_2 = \alpha^{R(i_1+i_2)} \bmod p$. Upon receiving L_2, Alice creates $L_3 = \alpha^{R(x+i_2)} \bmod p$ which she then transmits to Bob on the third leg. Bob has a prior knowledge of both R and i_2 thus he can recover the key sent over the channel.

Thus the set of equations that should be solved by an intruder tapping the channel can be presented as follows:

$$L_1 = \alpha^{Ri_1} \bmod p = m \tag{12.10}$$

$$L_2 = \alpha^{R(i_1+i_2)} \bmod p = m' \tag{12.11}$$

$$L_3 = \alpha^{R(i_2+x)} \bmod p = m'' \tag{12.12}$$

Since it's impossible to solve for four unknown variables with only three equations, the key exchange is secure.

12.4.2 Mode of Operation

We call $R_0[n]$ the initialization vector at iteration 0, cycle number n, $x_0[n]$ the bit value of the message being transferred, and $i_1^0[n]$ and $i_2^0[n]$ the values of the transformations (keys) at iteration 0, cycle number n, associated with the sender (Alice, 1) and receiver (Bob, 2), respectively. The initial length of R_0 is denoted by z, thus after z cycles R_0 will be updated to a new string of values R_1 of the same length as R_0.

At iteration 0, cycle number n, the messages transmitted at $L_{1,0}$, $L_{2,0}$, and $L_{3,0}$ respectively are:

$$L_{1,0} = \alpha^{R_0[n]\, i_1^0[n]} \bmod p \tag{12.13}$$

$$L_{2,0} = \alpha^{R_0[n]\left(i_1^0[n]+i_2^0[n]\right)} \bmod p \tag{12.14}$$

$$L_{3,0} = \alpha^{R_0[n]\left(X_0[n]+i_2^0[n]\right)} \bmod p \tag{12.15}$$

It should be noted that the key to be transmitted $K_0[n] = \alpha^{X_0[n]} \bmod p$ has been used only in the last leg. This approach can also be used in the set of equations where the initialization vector R_0 was not used. Potentially, this is an additional deterrent to a cryptanalyst.

At the next cycle, Alice will use a new transformation set $i_1^0[n+1]$ and Bob should use $i_2^0[n+1]$ and a next value in the string of the initialization vector $R_0[n+1]$ will be used. It is worth noting that Alice and Bob do not have any restrictions on the transformations associated with R, and it does not need to commute with $i_1^0[n+1]$ and $i_2^0[n+1]$. Furthermore, it is worth noting that Alice and Bob can use the bits of R in any desired order as long as there has been a previous agreement on the way of using them. The operation of the proposed approach is depicted in Fig. 12.5. A simulation of the proposed approach is presented in Table 12.3 for random i_1, i_2, and x, with $\alpha = 5$, $x_{E2} = 20$, and $p = 97$.

After the addition of an extra dimension to the multi-stage protocol proposed in the previous section it is possible to consider it as a continuously refreshing key since, at each stage of the protocol, a new secret variable is added in order to secure the outcome of the previous stage. An eavesdropper having simultaneous access to the three stages of the protocol will not be able to compute the value of the sent bit, since he/she will be faced with the problem of solving a system of three equations

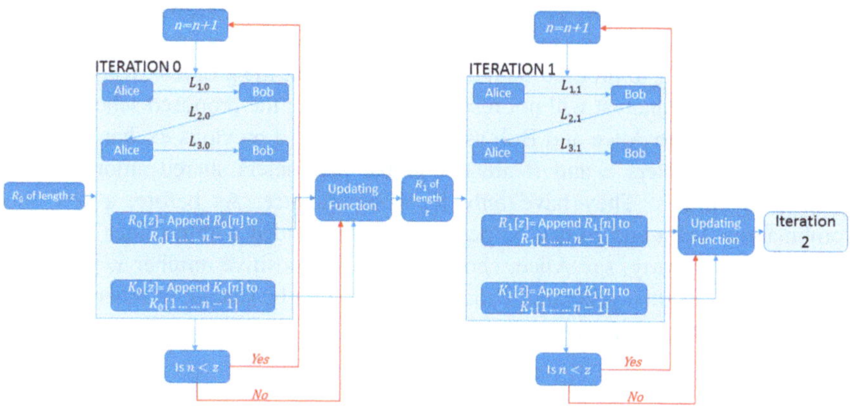

Fig. 12.5 The operation of the multi-stage protocol using four variables

Table 12.3 Simulation of the operation of the multi-stage protocol using an initialization vector R

	Cycle n	Key at Alice	i_1^0	i_2^0	$L_{1,0}$	$L_{2,0}$	$L_{3,0}$	Key at Bob	R_0
Iteration 0	1	14.28	2.41	5.75	45.32	70.91	32.92	14.28	4.71
	2	49.32	1.33	10.63	62.34	30	89	49.32	9.89
	3	24.55	2.87	10.05	41.5	8	33	24.55	29.95
	4	14.72	3.34	1.48	11.66	76.61	11.21	14.72	25.80
	5	62.82	4.07	3.49	39.42	37	87.11	62.82	27.50
	Cycle n	Key at Alice	i_1^1	i_2^1	$L_{1,1}$	$L_{2,1}$	$L_{3,1}$	Key at Bob	R_1
Iteration 1	1	59.92	3.19	4.91	65.59	67.31	94.26	69.92	64.94
	2	81.19	0.65	3.44	71.17	80.92	80.06	81.19	22.4
	3	27	4.66	6.16	42.20	40	69.78	27	207.9
	4	61.69	3.92	9.95	38.27	58	35	61.69	46.27
	5	88.65	2.95	8.82	78.46	78	41.18	88.65	8.44

with four variables. In addition, such an approach makes it impossible to launch a man in the middle attack.

The addition of an initialization vector to the three-stage protocol can be regarded as a door function to protect the message sent over the channel. Any illegitimate user is denied the ability of retrieving the value of the bit sent over the channel as long as he/she does not possess the value of the door function.

12.4.3 A Two-Stage Protocol

The protocol proposed in Sect. 12.4.2 can be further reduced into a two-stage protocol if one allows the communication to start at Bob's side and end at Bob's side as well (shown in Fig. 12.6). The protocol proceeds as follows: Alice and Bob initially share a large prime number p and its primitive root α and an initialization vector R of length z to proceed with the first iteration of the protocol. Knowledge of p and α by a cryptanalyst will not result in invalidating the proposed scheme, while R is only known to Alice and Bob as a remnant from a previous exchange. In any event, the parameters α and R are not global parameters shared among a large number of entities. They have only local significance. As before, a few such combinations (of α and p) can be permanently embedded within router pairs. The key exchanging party, say Alice, chooses a random positive number x, where x is less than p. Alice then generates $\alpha^x \bmod p$ as the key to be shared. In other words, as stated earlier, the key K can be represented as,

$$K = \alpha^x \bmod p \tag{12.16}$$

As discussed before, the key at the first iteration is denoted by $K[n]_0 = \alpha^{X_0[n]} \bmod p$. Bob starts the communication by sending $L_1 = \alpha^{R\, i_2} \bmod p$, where i_2 a random integer and is known only to Bob. Upon receiving the message, Alice encodes $\alpha^x \bmod p$, which is the key to be transmitted, and protects the message using the initialization vector R, and sends $L_2 = \alpha^{R(x+i_2)} \bmod p$ to Bob. Bob receives the message his prior knowledge of R and i_2 gives him the ability to decode the message sent from Alice.

The initialization vector R is updated periodically to insure a fully secure information transfer between Alice and Bob.

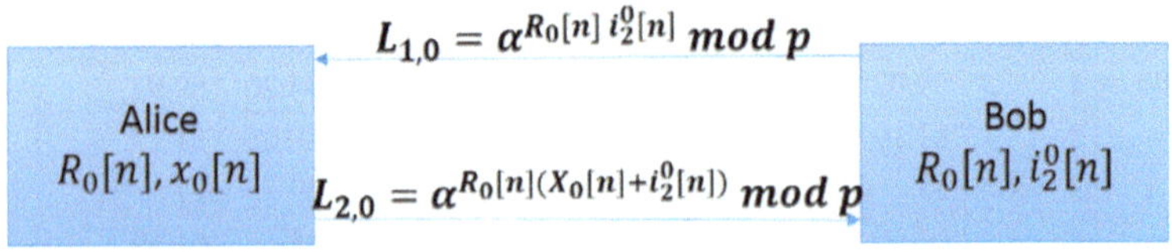

Fig. 12.6 Key exchange scheme using the two-stage protocol (iteration zero cycle n)

$R_0[n]$ is the initialization vector at iteration 0 cycle number n, $x_0[n]$ the bit value of the message being transferred, $i_2^0[n]$ the values of the transformations (keys) at iteration 0 cycle number n of the receiver (Bob). The initial length of R_0 is denoted by z, thus after z cycles R_0 will be updated to a new string of values R_1, of the same length.

At iteration 0 cycle number n the messages transmitted at $L_{1,0}$, and $L_{2,0}$ are:

$$L_{1,0} = \alpha^{R_0[n]\, i_2^0[n]} \bmod p \tag{12.17}$$

$$L_{2,0} = \alpha^{R_0[n]\left(X_0[n] + i_2^0[n]\right)} \bmod p \tag{12.18}$$

At the next cycle Bob will use $i_2^0[n+1]$ and the next value in the string of the initialization vector $R_0[n+1]$ will be used. It is worth noting that Alice and Bob do not have any restrictions on the transformations associated with R, it can be an addition to the exponent, a multiplication or any other mathematical transformation.

At the end of each iteration an updating function having its inputs as the randomly generated key K and the initialization vector R is used to generate a new initialization vector to be used in the next iteration. As shown in Fig. 12.7, R_0 is used to generate the key at iteration 0. Then, using R_0 and K_0 the updating function generates a new initialization vector R_1 of length z to be used for iteration 1.

12.4.4 Braiding Concept

The concept of using different versions of the multi-stage protocol in the same communication called the braiding function was introduced in [19], and is shown in Fig. 12.8. Braiding is the concept of sharing keys between two parties using

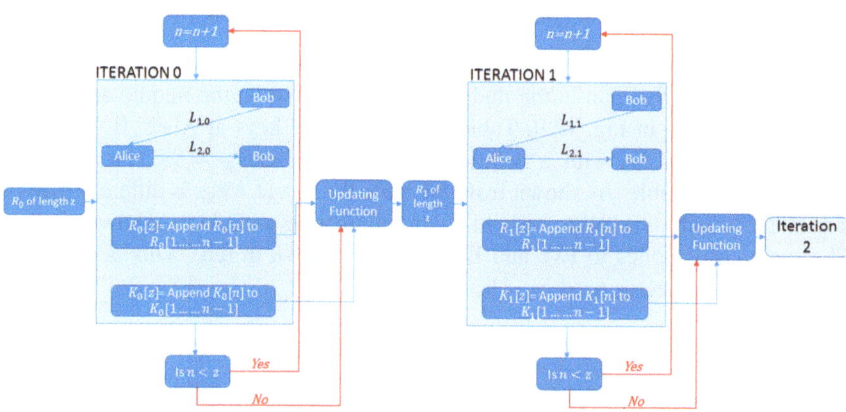

Fig. 12.7 The operation of the multi-stage protocol using three variables

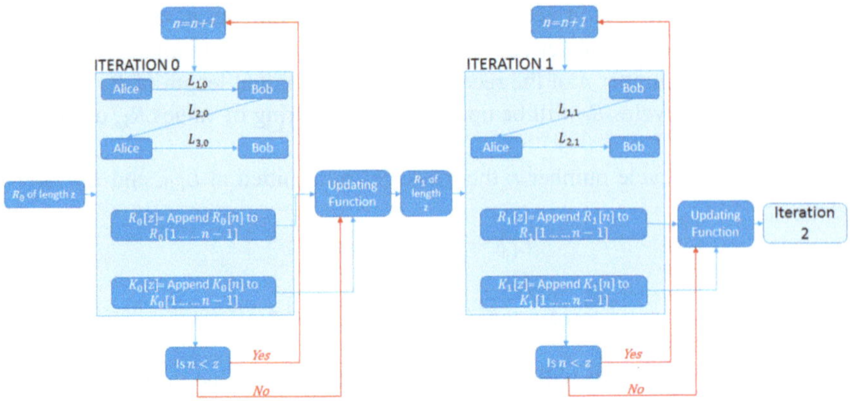

Fig. 12.8 The operation of the braided multi-stage protocol

different number of stages, m. Alice and Bob can agree on how many stages to use at the onset of the key distribution process. The variability associated with m (which can be changed at will by the communicating parties) will be an additional means of frustrating the eavesdropper Eve.

In Fig. 12.8, the braiding concept of multi-stage protocols is shown. Given an initialization vector R_0 of length z Alice and Bob execute a multi-stage protocol with $m = 3$ for z cycles. Then after sharing a key K_0 of length z, the initialization vector R_0 is updated into R_1 using an updating function. The new initialization vector R_1 is based on both K_0 and R_0. R_1 is now used to do the next iteration of the key distribution process. At this iteration Alice and Bob use $m = 2$ to share K_1.

12.4.5 Man in the Middle Attack on a Multi-stage Protocol Using an Initialization Vector

In this section, we show that a multi-stage protocol using an initialization vector is not vulnerable to the man in the middle attack. The man in the middle attack stage protocol is shown in Fig. 12.9. Table 12.4 shows the keys at Alice, Bob and Eve for i_1, i_2, i_{E1}, and i_{E2} with $\alpha = 5, x = 15$, $x_{E2} = 20$, and $p = 97$, the simulation presented in this table. As shown from the table the key at Alice is different than the key that Eve will attempt to use while communicating with Alice (shown in blue). The same case applies to Eve and Bob as well (shown in red). This is due to the initialization R_0 vector used by Alice and Bob.

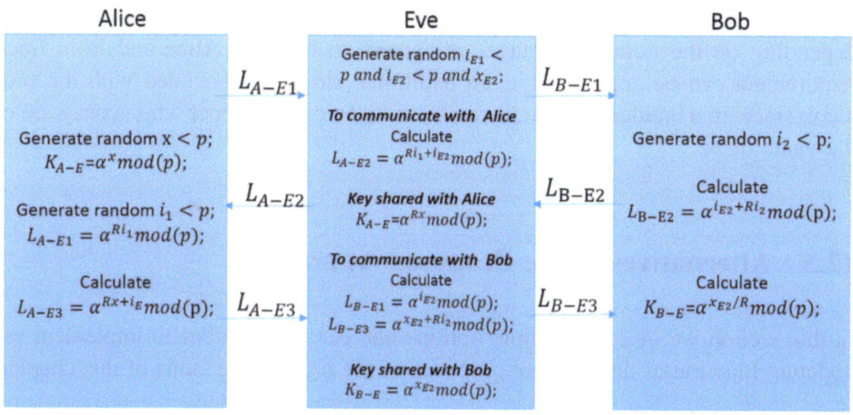

Fig. 12.9 Man in the middle attack on the multi-stage using an initialization vector

Table 12.4 Man in the middle attack to a multi-stage protocol using an initialization vector

Cycle	Key at Alice	L_{A-E1}	L_{A-E2}	L_{A-E3}	Key Alice-Eve	Key Eve-Bob	L_{B-E1}	L_{B-E2}	L_{B-E3}	Key at Bob
1	46	44.42	93	61	63.12	61	42	93	86.87	93
2	46	36.50	5	37	75.29	80.62	33	68	70	93
3	46	15.78	33	61	50.01	45.75	46	34	91	93
4	46	81.95	16	52.24	33.19	60.34	36	46	46	93
5	46	69.21	26	54	4.04	33.81	35	8	22	93

12.4.6 Characteristics of the Proposed Protocol

The protocol proposed in this chapter is a multi-stage ultra-secure router-to-router spontaneous key exchange system. This protocol represents a decentralized approach to conventional key distribution schemes and is meant for use on a link-by-link basis between two communicating parties. In other words, the overhead of sharing public keys throughout the network is obviated as well as the need of a third party to distribute secret keys between the nodes of the network.

The proposed protocol is based on the usage of random number generators. Two random number generators at both Alice's and Bob's side are needed to generate i_1 and i_2, respectively. Furthermore, the security of the proposed approach against a man in the middle attach is guaranteed as long as an initialization vector is already in the possession of the communicating parties at the time communication first starts. This initialization vector can be embedded in each of the two communicating router pairs by the manufacturer, and then updated as often as chosen by the communicating routers.

Finally, the multi-stage protocol requires several transmissions on the channel depending on the number of stages m agreed on between Alice and Bob. Such requirement can be considered as an additional strength associated with the protocol, since, in a braided scheme, it can frustrate an eavesdropper who is unaware of the value of m.

12.5 Alternatives to the Proposed Approach

In this section we discuss multiple options one can use in order to implement the updating function of the protocol proposed in the previous sections of this chapter.

12.5.1 Alternative I—RSA

The RSA algorithm has the advantage of being one of most widely used and studied encryption methods today, and is extremely elegant, simple, and well-tested. As the default algorithm used by many SSL providers, as well as the basic public key encryption scheme most others are compared to, RSA is used here as a baseline for the comparison of other encryption methods, even though it is not as storage or processing-efficient as other algorithms studied, and requires the use of longer key lengths for equivalent security. Current commonly used RSA key lengths include 1024 and 2048 bits.

The basic principle of RSA security rests on the theory that it is extremely difficult to factor the product of two large prime numbers into its constituent factors. Each individual in the RSA network must create two complimentary keys, commonly referred to as a public key and a private key, with each key able to decrypt messages enciphered using its compliment. To create this keypair, Alice and Bob much each do the following [3].

Choose two similar large prime numbers p and q, which are within a few digits of each other in length. P and q are multiplied together to form a modulus n.

An integer e is chosen such that $e<(\phi(n))$ and e and $\phi(n)$ are coprime $(\phi(n)=n-p-q+1)$. A common value of e is $65,537\,(216+1)$. The public key consists of n and e (modulus and public key exponent).

The modular multiplicative inverse of e mod($\phi(n)$) $(d\equiv e^{-1}(\mathrm{mod}(\phi(n))))$ is calculated, and the private key consists of n and d (modulus and private key exponent).

Message encryption may then be expressed, using the one key, as $C=(m^{K1})(\mathrm{mod}\,n)$, while decryption uses the other key as $P=C^{K2}(\mathrm{mod}\,n)$.

Typically, as the sending party must know the recipient's public key, as well as their own private key, RSA is not used within a self-contained system. Key generation for large primes may also be time consuming and resource intensive. Instead, third-party organizations typically create and distribute key pairs for a wide range of customers, and must be trusted to verify that a given public key corresponds to the stated owner's private key. Issued keypairs are generally valid for a set length of time, after which a new key must be requested via the central distribution authority.

As our proposed router system must be self-contained after initial manufacture, this third party distribution method is not feasible, and we cannot rely on external communication for the generation and distribution of additional certificates, requiring a slight modification of the standard RSA system. Instead, Alice's router will need to be initialized with pre-stored values for Alice's private key and Bob's public key, and Bob's will have Alice's public key and his own private key. In this scenario, it is not necessary for either party to know their own public key, and all 4 keys are kept private within the network.

Encryption and decryption function as standard RSA operations, with Alice encrypting data with Bob's public key, and Bob decrypting with his private key, and vice versa. After a data threshold is exceeded, Alice and Bob will both calculate new RSA keypairs, and encrypt and send their new public keys using the old keys. For example, Alice's new public key would be encrypted first with her old private key for authentication, then with Bob's old public key for security, then sent to Bob. Bob would decrypt using his own old private key, then Alice's old public key. Once both parties have received the new keys, all data will be transmitted using these. This system would allow for the use of RSA indefinitely, with rapid key updates, without the necessity of a third party. In the event of a communication failure due to data loss or malicious action, it may be necessary to switch to a new pre-shared certificate pair and begin the process again.

Storage requirements for an n-bit RSA system are comparatively large, as larger key lengths are needed to assure equivalent security. Specifically, each router using this n-bit RSA algorithm will need to store one public and one private key, each consisting of an n-bit modulus and a smaller exponent (also of maximum length about n) for maximum total requirement of $4n$ bits per router. Processing time for RSA is also comparatively long, due to the larger key lengths and exponentiation operations required. The security of RSA is based upon the difficulty of the factorization problem. As with the discrete logarithm attack, the current approach to integer factorization involves the general number field sieve algorithm [20], which for an integer n will arrive at a solution in $L_p[1/3; 1.923]$ i.e. $\left(e^{(1.923+O(1))(\log n)^{1/3}(\log\log n)^{2/3}}\right)$. A quantum computer, should it ever exist, may factor large integers in polynomial time [21].

Although it is obvious that RSA offers several disadvantages when compared to other symmetric and asymmetric ciphers, it also offers at least one key advantage when compared to the other algorithms herein: message authentication. Unlike discrete logarithm, ECC, or AES encryption, since neither Alice nor Bob know the other individual's private key, it would be possible for a third party external audit, given hardware access to both router keys and all traffic sent, to determine the sender of all encrypted data. Using the other encryption systems, given the encrypted data alone, it is possible to determine that either Alice or Bob sent a message, but not to authenticate which one encrypted the data.

12.5.2 Alternative II—AES

AES, based upon the Rijndael cipher, was announced by the National Institute of Standards and Technology in 2001, and was shortly thereafter approved as an accepted encryption standard by the United States Federal government. AES, similar to its predecessor, DES, is a symmetric block cipher, using a shared secret key to encrypt a data stream one block at a time. In AES, each 128-bit data block undergoes 10–14 rounds (depending on key length) of permutations, and substitutions, and additions [3]. AES is an extensively used and studied algorithm, and like most symmetric ciphers, offers advantages in terms of required processing power, processing time, and key length when compared to asymmetric ciphers such as RSA and ECC. The simplicity of each round enables simple and rapid implementation on any 8-bit processor, while the chaining of multiple rounds per block provides excellent security. Many modern processors implement AES optimization instructions, making this an excellent algorithm choice for use with existing hardware.

To modify AES for use in our closed system, Alice and Bob's routers will both require a single pre-shared AES key, and a reliable PRNG. Initial communication will be made using the pre-shared key. After a data threshold has been reached, similar to the discrete logarithm system, Alice and Bob will input the decrypted data into an algorithm (such as a cryptographic hash function) to generate a random value R. This value will be used as a PRNG seed on both systems to generate identical intermediate keys of the desired AES key length. To compensate for any bias in the data used to generate R (similar data between datacycles may lead to a smaller PRNG seedpool), the intermediate key may be XOR'd with the previous AES key to generate a new, random shared secret key by which further communication will be encrypted.

As mentioned earlier AES offers efficient processing time, and the storage requirements for this system are minimal, requiring a single pre-shared key to be saved on each router, much shorter than a security-equivalent RSA keypair. No effective attacks are currently known against AES, with the current best attacks only a few orders of magnitude above the worst-case brute force scenario, and requiring infeasibly large amounts of storage space [21]. Unlike asymmetric encryption algorithms, AES is resistant to attacks by theoretical future quantum computers. In the event of a communication failure due to data loss or malicious action, it may be necessary to switch to a new pre-shared key and begin the process again.

12.5.3 Alternative III—ECC

Elliptic curve cryptography (ECC) is an asymmetric cryptographic system, which uses a variant of the discrete logarithm problem as applied to points in an elliptic curve group as the core of its security. Many consumers have recently begun adopting ECC as an alternative to RSA, due to its efficiency in both key size and processing requirements. Careful choice of the ECC curve is necessary to avoid potential security hazards.

In elliptic cryptography, first a curve is chosen, with variables and coefficients restricted over either the finite field GF($2m$) of the form $y^3 + xy = x^3 + ax^2 + b$ or a prime curve over Z_p and modulo p where variables and coefficients range from 0 to $(p-1)$ of the form $y^2\text{mod}(p) = (x^3 + ax + b)\text{mod}(p)$.

In the prime curve case, there are a limited number of non-negative integer points between $(0,0)$ and $(p-1, p-1)$ which satisfy any given elliptic curve values for a and b. Similarly, for the finite field case, there will be a limited number of (x, y) integer values that lie on the curve for any given values of a and b.

These points are used to define a finite abelian group, with rules for addition defined specifically for the abelian group, similar to modular multiplication in conventional algorithms. Likewise, multiple additions are preformed similarly to modular exponentiation. Using Abelian group rules, given two points M and N, $M = kN$ is easily calculated given k and N, but difficult to calculate given M and N, forming the one-way trapdoor function at the basis of elliptic cryptography.

Generally, the curve parameter values of a, b and z, C and n are made public, and often correspond to one of several well-studied elliptic curves. a and b are the coefficients discussed earlier, forming the curve $E_z(a, b)$, where z is an integer in the finite field $2m$ (finite field curve) or a large prime number (prime curve). A base point C is picked such that the smallest positive integer n that satisfies $nC = 0$ is very large. With all curve parameters defined, Alice and Bob may begin the key selection process [3].

(1) Alice and Bob both choose secret integers I_A and I_B less than n as their private keys.
(2) Public keys are generated according to $P_A = I_A \times C$ and $P_B = I_B \times C$ and shared with each other.
(3) Generate a common secret key by multiplying the known private key with the opposite public key, with $I_A \times P_B = I_B \times P_A$
(4) To encrypt or decrypt data, the data is first encoded as a point M on the elliptic curve, and then sent as a ciphertext message as a pair of points $(kC, M + kP)$ with k as any chosen positive integer, and decrypted with the matching private key using $(M_x I - M_y)$

Modifying this system to function in our self-contained router environment involves a process similar to that used for RSA. All curve parameters are assumed to be publicly known, and use of a known secure curve is assumed. Each router must be initialized with secret data corresponding to its own private key, and the public key of the other router. Again, it is not strictly necessary for each party to know or retain its own public key, and in any case, all four key values are kept secret within the network.

Encryption and decryption function as standard ECC operations, with Alice encrypting data with Bob's public key, and Bob decrypting with his private key, and vice versa. After a data threshold is exceeded, Alice and Bob will both calculate new public and private ECC keys, choosing new secret integers, and encrypt and send their each other their new public keys using their old private keys. Once both parties have received the new keys, all data will be transmitted using these. This system would allow for the use of ECC indefinitely, with rapid key updates, without the necessity of a third party. In the event of a communication failure due to data loss or malicious action, it may be necessary to switch to a new pre-shared certificate pair and begin the process again. Unlike in RSA, the use of a common secret key prevents message authentication via external audit.

Storage requirements for ECC involve two large integers of size n or smaller, corresponding to the public and private keys, on each router, for a total maximum storage capacity of $2n$ per shared secret per router. Key lengths used are much shorter than those needed for equivalent RSA or discrete logarithm security levels, about double the size of that found in symmetric encryption systems. Likewise, while not quite as processing-efficient as a symmetric cryptosystem, ECC offers large performance gains when compared to RSA. The best known attack to ECC is Pollard's Rho [22] which may be run via full parallelization across multiple independent threads, processors, or even locations, with occasional communication via a central server. Pollard's Rho needs relatively little memory, but is nevertheless not computationally feasible for currently used ECC curve parameters, even accounting for massive parallelization. As with other public key protocols, ECC is expected to be vulnerable to attack by quantum computers, once such exist.

12.6 Summary

This chapter has presented an ultra-secure router-to-router key exchange system using multi-stage protocols. The key exchange process can be initiated by either party at will and can be carried out as often as necessary. The proposed system is based on the use of discrete logarithms. Alternatively Elliptic Curve Cryptography can be used as well. The main cryptographic strength of the proposed protocols lies in the use of multi-stage transmission where the number of variables exceeds the number of stages by one, ensuring that the number of possible measurements is one less that the number of variables. This makes the key transfer secure. Furthermore, since the keys can be exchanged as often as necessary at the speed of communication, one can stipulate that the security of the proposed system approaches that of contemporary QKD systems. In other words, the level of security attainable by the proposed system is comparable to QKD systems but without the baggage of limited distance and low speed associated with the latter. Recent literature in quantum-secure communication in a multi photon environment [1, 7, 18, 19] has suggested that the multi-photon approach can offer quantum level security obviating the need for single photon generators and thus removing the speed and distance barrier caused by single photons. This chapter has shown that a cryptographic strength similar to the multi-photon approach is attainable in electronics as well.

The proposed technique has used discrete logarithms as the basis for secure communication. Other possible techniques were proposed in this chapter. These techniques are based on the use of RSA, AES, and ECC.

Ultimately, algorithm choice will likely be determined by system needs, and the availability of supporting hardware. While advances in modern solid-state storage make it unlikely that secret storage space is ever a practical limitation of the proposed router-to-router key exchange system, algorithm processing efficiency, data transfer efficiency, and key generation time may have a much larger impact on system design.

If authentication is needed, RSA, the weakest algorithm in terms of key generation and processing efficiency, is the clear choice. If, however, authentication is not needed, then symmetric key systems, such as the AES, offer the most efficient alternative. Alternatively, a hybrid of both techniques may be used, offering on-demand authentication and efficient non-authenticated communication, as necessary.

As a final consideration, any encryption system is vulnerable to physical hardware compromise. If an attacker can gain access to the shared secret data stored on the router's security hardware, even the most secure encryption framework will be compromised. Therefore, care must be taken during hardware manufacturing and distribution to ensure that these keys are not copied or prematurely accessed.

References

1. El Rifai, M., & Verma, P. K. (2013). An algorithmic approach to securing the three-stage quantum cryptography protocol. In *2013 12th IEEE international conference on trust, security and privacy in computing and communications (TrustCom)*, pp. 1803–1807: IEEE.
2. Parmar, N. J., & Verma, P. K. (2017). A comparative evaluation of algorithms in the implementation of an ultra-secure router-to-router key exchange system. In *Security and communication networks* (vol. 2017, Article ID 1467614, 7 pages). https://doi.org/10.1155/2017/1467614.
3. Stallings, W., & Tahiliani, M. P. (2014). *Cryptography and network security: Principles and practice*. London: Pearson.
4. Bennett, C. H., & Brassard, G. (2014). Quantum cryptography: Public key distribution and coin tossing. *Theoretical Computer Science, 560,* 7–11.
5. Scarani, V., Acin, A., Ribordy, G., & Gisin, N. (2004). Quantum cryptography protocols robust against photon number splitting attacks for weak laser pulse implementations. *Physical Review Letters, 92*(5), 057901.
6. Bennett, C. H. (1992). Quantum cryptography using any two nonorthogonal states. *Physical Review Letters, 68*(21), 3121.

7. Mandal, S., et al. (2012). Implementation of secure quantum protocol using multiple photons for communication. *arXiv preprint*. arXiv:1208.6198.
8. Zhao, Z., Yang, T., Chen, Y.-A., Zhang, A.-N., & Pan, J.-W. (2003). Experimental realization of entanglement concentration and a quantum repeater. *Physical Review Letters, 90*(20), 207901.
9. Menn, J. (2012, 2 Feb). Key internet operator VeriSign hit by hackers. *Reuters*.
10. Ore, O. (2017). *Invitation to number theory*. American Math Soc.
11. Rosen, K. H. (1993). *Elementary number theory and its applications*. Addison-Wesley.
12. Frisch, M., & Simmons, G. J. (1992). *Public-key cryptography: State of the art and future directions: EISS workshop*, Oberwolfach, Germany, 3–6 July 1991. Springer.
13. Diffie, W., & Hellman, M. (1976). New directions in cryptography. *IEEE Transactions on Information Theory, 22*(6), 644–654.
14. Diffie, W., & Hellman, M. E. (1977). Special feature exhaustive cryptanalysis of the NBS data encryption standard. *Computer, 10*(6), 74–84.
15. Merkle, R. C., & Charles, R. (1979). Secrecy, authentication, and public key systems.
16. Needham, R. M., & Schroeder, M. D. (1978). Using encryption for authentication in large networks of computers. *Communications of the ACM, 21*(12), 993–999.
17. Kak, S. (2006). A three-stage quantum cryptography protocol. *Foundations of Physics Letters, 19*(3), 293–296.
18. Chen, Y., Kak, S., Verma, P. K., Macdonald, G., El Rifai, M., & Punekar, N. (2013). Multi-photon tolerant secure quantum communication—From theory to practice. In *2013 IEEE international conference on communications (ICC)*, pp. 2111–2116: IEEE.
19. Darunkar, B., & Verma, P. K. (2014). The braided single-stage protocol for quantum secure communication. In *Quantum information and computation XII* (vol. 9123, p. 912308). International Society for Optics and Photonics.
20. Coppersmith, D. (1993). Modifications to the number field sieve. *Journal of Cryptology, 6*(3), 169–180.
21. Shor, P. W. (1999). Polynomial-time algorithms for prime factorization and discrete logarithms on a quantum computer. *SIAM Review, 41*(2), 303–332.
22. Bos, J., Kaihara, M., Kleinjung, T., Lenstra, A. K., & Montgomery, P. L. (2009). *On the security of 1024-bit RSA and 160-bit elliptic curve cryptography*.

FSC
www.fsc.org
MIX
Papier aus verantwortungsvollen Quellen
Paper from responsible sources
FSC® C105338

If you have any concerns about our products,
you can contact us on
ProductSafety@springernature.com

In case Publisher is established outside the EU,
the EU authorized representative is:
Springer Nature Customer Service Center GmbH
Europaplatz 3, 69115 Heidelberg, Germany

Printed by Libri Plureos GmbH
in Hamburg, Germany